Experientia Supplementum

Volume 110

AF817967

More information about this series at http://www.springer.com/series/4822

Roderick A. Slavcev • Shawn Wettig •
Zhiheng Zeng
Editors

Nanomedicine

Gene Delivery, Imaging and Evaluation Systems

Springer

Editors
Roderick A. Slavcev
School of Pharmacy
University of Waterloo
Waterloo, Ontario, Canada

Shawn Wettig
School of Pharmacy
University of Waterloo
Waterloo, Ontario, Canada

Zhiheng Zeng
School of Pharmacy
University of Waterloo
Waterloo, Ontario, Canada

ISSN 1664-431X ISSN 2504-3692 (electronic)
Experientia Supplementum
ISBN 978-3-030-08658-9 ISBN 978-3-319-78259-1 (eBook)
https://doi.org/10.1007/978-3-319-78259-1

© Springer International Publishing AG, part of Springer Nature 2018
Softcover re-print of the Hardcover 1st edition 2018
This work is subject to copyright. All rights are reserved by the Publisher, whether the whole or part of the material is concerned, specifically the rights of translation, reprinting, reuse of illustrations, recitation, broadcasting, reproduction on microfilms or in any other physical way, and transmission or information storage and retrieval, electronic adaptation, computer software, or by similar or dissimilar methodology now known or hereafter developed.
The use of general descriptive names, registered names, trademarks, service marks, etc. in this publication does not imply, even in the absence of a specific statement, that such names are exempt from the relevant protective laws and regulations and therefore free for general use.
The publisher, the authors and the editors are safe to assume that the advice and information in this book are believed to be true and accurate at the date of publication. Neither the publisher nor the authors or the editors give a warranty, express or implied, with respect to the material contained herein or for any errors or omissions that may have been made. The publisher remains neutral with regard to jurisdictional claims in published maps and institutional affiliations.

Printed on acid-free paper

This Springer imprint is published by the registered company Springer International Publishing AG part of Springer Nature.
The registered company address is: Gewerbestrasse 11, 6330 Cham, Switzerland

Contents

Chapter 1
Introduction

Roderick Adrian Slavcev

Gene therapy is the delivery of nucleic acid for the expression of a therapeutic product in order to treat diseases on a genetic level. This is especially well suited for diseases that involve missing, defective, or overexpressing genes. There are two primary strategies for transgene delivery: viral and non-viral. Gene therapy clinical trials (Edelstein 2017) show greater emphasis on viral vector-based antigen, cytokine, tumor suppressor, deficiency, and suicide transgene delivery (http://www.abedia.com/wiley/genes.php). Although viral vectors are efficient in gene delivery, alternatives to viruses have been explored due to some of their major limitations, which include potential for oncogenesis due to chromosomal integration or activation of oncogenes/inactivation of oncogene regulators, limited capacity for transgenic materials (i.e., size of gene or genes to be transferred), the generation of infectious viruses due to recombination, and concerns regarding production and storage (Al-Dosari and Gao 2009). This book aims to cover a wide spectrum in regard to systems specific for gene delivery, imaging, and evaluation methodologies. Under gene delivery, non-viral delivery systems, both chemical and physical, are discussed: microinjection, electroporation, gene ballistic, lipopolyfection, magnetofection, sonoporation, hyperthermia, magnetic gold, and carbon-based nanoparticle methods. Technological advancements in molecular imaging will be covered, including multiphoton microscopy, surface-modified quantum dots, positron-emission tomography, magnetic resonance imaging, spectroscopic techniques, fluorescence resonance energy transfer microscopy, fluorescence correlation spectroscopy, fluorescence cross correlation spectroscopy, and fluorescence recovery after photobleaching. The advent of genetically engineered cells with fluorescent proteins has made these techniques a common tool for studying molecular dynamics in living cells (Giepmans et al. 2006). These imaging modalities, which have opened new doors, giving new cellular and molecular insight, are

R. A. Slavcev (✉)
University of Waterloo, Waterloo, ON, Canada
e-mail: slavcev@uwaterloo.ca

© Springer International Publishing AG, part of Springer Nature 2018

R. A. Slavcev et al. (eds.), *Nanomedicine*, Experientia Supplementum 110,
https://doi.org/10.1007/978-3-319-78259-1_1

discussed in greater detail in the book. In addition, real-time imaging and in vivo analysis of reporter gene expression give the ability to monitor spatiotemporal aspects of gene delivery in a noninvasive manner. It is now possible to use fluorescent/bioluminescent genetic reporter strategies to evaluate the pharmacokinetic profiles, biodistribution, and therapeutic efficacy of gene/vector of interest in live animals (Giepmans et al. 2006; Bai and Bornhop 2012). Finally, this book touches upon the common methodologies involved in evaluating transgene expression at transcription and translational levels such as reverse transcription polymerase chain reaction, serial analysis of gene expression, transcriptome array northern/western blotting, enzyme-linked immunosorbent assay, and proteome array. Overall, we aim to provide a holistic view of gene therapy from advancements in vector design to new emerging imaging and evaluation modalities.

References

Al-Dosari MS, Gao X (2009) Nonviral gene delivery: principle, limitations, and recent progress. AAPS J 11:671–681. https://doi.org/10.1208/s12248-009-9143-y

Bai M, Bornhop DJ (2012) Recent advances in receptor-targeted fluorescent probes for in vivo cancer imaging. Curr Med Chem 19:4742–4758

Edelstein M (2017) Gene therapy clinical trials worldwide. J Gene Med. http://www.wiley.com/legacy/wileychi/genmed/clinical/

Giepmans BNG, Adams SR, Ellisman MH, Tsien RY (2006) The fluorescent toolbox for assessing protein location and function. Science (80–) 312:217–224. doi:https://doi.org/10.1126/science.1124618

Chapter 2
Non-viral Gene Delivery

Chi Hong Sum, Samantha Marisha Shortall, Shirley Wong, and Shawn David Wettig

Contents

Abstract Although viral vectors comprise the majority of gene delivery vectors, their various safety, production, and other practical concerns have left a research gap to be addressed. The non-viral vector space encompasses a growing variety of physical and chemical methods capable of gene delivery into the nuclei of target cells. Major physical methods described in this chapter are microinjection, electroporation, and ballistic injection, magnetofection, sonoporation, optical transfection, and localized hyperthermia. Major chemical methods described in this chapter are lipofection, polyfection, gold complexation, and carbon-based methods. Combination approaches to improve transfection efficiency or reduce immunological

C. H. Sum · S. M. Shortall · S. Wong · S. D. Wettig (✉)
University of Waterloo, School of Pharmacy, Waterloo, ON, Canada
e-mail: wettig@uwaterloo.ca

© Springer International Publishing AG, part of Springer Nature 2018

R. A. Slavcev et al. (eds.), *Nanomedicine*, Experientia Supplementum 110,
https://doi.org/10.1007/978-3-319-78259-1_2

response have shown great promise in expanding the scope of non-viral gene delivery.

Keywords Non-viral gene delivery · Microinjection · Magnetofection · Lipofection · Polyfection · Gold nanoparticles · Carbon-based gene delivery · Optical transfection · Sonoporation · Localized hyperthermia

2.1 Introduction

Viral vectors remain the most widely studied gene delivery vector for gene therapy applications, due primarily to the high delivery efficiencies inherent to these vectors; however, the various safety (Raper et al. 2003; Hacein-Bey Abina et al. 2003, 2008) and other, practical, concerns associated with viral vectors continue to drive research into non-viral gene delivery methods. Non-viral methods can be broadly classified into physical- and chemical-based methods. Chemical-based vectors are generally comprised of materials that are substantially less toxic and/or less immunogenic compared to viral vectors and, through the use of targeting ligands, can be targeted to specific cell populations, while physical methods allow for spatial precision (i.e., site-specific delivery of DNA) (Al-Dosari and Gao 2009). Major chemical methods of gene delivery will be described in detail in sections below, along with select physical/combined methods, specifically those that have and continue to generate a significant number of research publications. Other methods, such as electroporation, direct injection, microinjection, and other injection-based systems, remain under investigation, in some cases in ongoing clinical trials; the drawback with all of these methods is the invasive nature of the techniques. This can be balanced by, in some cases, the very high effectiveness of the localized injection of DNA into the target tissue(s) bypassing systemic and intracellular barriers.

2.2 Physical Methods

In the direct injection method, first reported in 1990 (Wolff et al. 1990), DNA uptake occurs primarily in the area of the needle track, pointing to the importance of physical damage to the tissue (Al-Dosari and Gao 2009). The majority of ongoing clinical trials involving direct administration of naked nucleic acid (427 in total as of 2016) are directed to cancer or cardiovascular disease (http://www.wiley.com/legacy/wileychi/genmed/clinical/). While the direct injection method is simple, with no concerns regarding vector toxicity, very low gene expression is observed as a result of the various barriers that impede cellular entry and expression (such as the negatively charged cell membrane, intra- and extracellular endonucleases, aggregation by serum proteins leading to clearance by the reticuloendothelial system, etc.). Microinjection, a technique originally developed for manipulation and transfer of material into living cells (Feramisco et al. 1999), allows for injection of genetic material directly into the nucleus via a micropipette. While microinjection gained a great deal of popularity in the early 1980s, it did not demonstrate efficiencies as high as those observed with viral vectors, is time-intensive (only a certain number of cells can be injected in a given amount of time), and requires a high degree of technical skill on the part of the operator (Cline 1985). Advances in manufacturing capabilities now allow for the production of microneedle patches, which enables dermal and transdermal delivery of drugs in general and DNA (Mellott et al. 2013) in particular [see a number of excellent reviews by Prausnitz (2004) and Kim et al. (2012)]; however, the method still suffers from a number of disadvantages, including primarily localized delivery, as well as imprecise and/or small dose limitations, and a high degree of complexity with respect to the manufacture of the patches themselves (van der Maaden et al. 2012).

Electroporation is perhaps one of the most effective and extensively studied and used methods for non-viral DNA transfection, and Mellott et al. provide a recent and extensive review of the technique (Mellott et al. 2013). The method utilizes electrical pulses to permeabilize cellular membranes (creating pores hence the term electroporation) allowing for the introduction of DNA into the cells and was first introduced by Neumann et al. (1982). While the method has been extremely successful both in vitro and in vivo, in a wide variety of cell lines and/or tissues, the most significant disadvantage is that the method is highly complex, requiring careful determination of physical and biological parameters for each application, as well as a skilled technician for its administration. This, coupled with the invasive nature needed for most applications (electrodes must be in contact with the tissue of interest), has limited its application, particularly in clinical settings. Improvements in the technology has led to ongoing clinical trials beginning in 2004 (Gothelf and Gehl 2010; Gehl 2014; Heller and Heller 2015), which show the high safety, tolerability, and efficacy of gene electrotransfer, particularly to the skin. Recent advances on gene electrotransfer to the skin or muscle are particularly promising, due to the accessibility of the tissues. Donate et al. observed that increased temperature upon electroporation improved DNA uptake, expression duration, and reduced tissue damage as a consequence of lower voltage requirements (Donate et al. 2016).

The final physical method of DNA transfection to be introduced is that of ballistic injection. Ballistic injection (also known as the "gene gun") differs from direct or microinjection in that it is a needle-free injection method. Instead, plasmid DNA is combined with metal nano- or microparticles (typically gold) and "fired" as a projectile from a pressurized ballistic device (hence the term "gene gun"). The technique was first introduced in 1987 (Sanford et al. 1987) and has gained popularity for applications such as DNA vaccines due to the ability of the technique to readily deliver DNA across the stratum corneum deeper into the epidermis of the skin (Wells 2004), but further application has been limited due to the inability to reach deeper tissues, in vivo, although there are numerous examples of successful delivery to the skeletal muscle, heart, tumor, and embryonic tissues in murine or rat models (see Mellott et al. (2013) and Wells (2004) for reviews on the ballistic method). Additional limitations are inflammation and potential tissue damage and a lack of specificity with both targeted and nontargeted cells being transfected (Mellott et al. 2013).

From the above, it is clear that there is not only great potential for the physical methods identified above, but they have also had excellent success, particularly in in vivo applications. In all cases, the limited clinical application of these methods results from the technical complexity of the method, the invasive nature of the method, or even a combination of both factors. As will be seen in the next section, chemical-based non-viral transfection methods offer numerous solutions to both the technical complexity and the invasive nature of some of the physical methods while still maintaining reasonable or even high levels of transfection efficiency.

2.3 Lipofection and Polyfection

Cationic lipid- and polymer-based carriers make up the most common non-viral gene delivery vectors. The first lipid-mediated delivery of nucleic acid was reported in the 1980s (Fraley et al. 1980; Felgner et al. 1987; Wu and Wu 1987). These systems were and are still considered highly promising gene carriers based on their ability to interact with both the anionic hydrophilic backbone of nucleic acids and with the anionic hydrophobic plasma membrane. Significant progress has since been made in characterizing and improving the interactions between both lipid/DNA (lipoplex) and polymer/DNA (polyplex) complexes and cellular barriers.

While they comprise an incredibly diverse vector space, cationic lipids do share an overall structure, such as (1) a cationic head group, (2) hydrophobic tail domains, and (3) a linker between these groups. Electrostatic interactions between the positively charged lipid head groups or polymer molecules with the negatively charged DNA leads to the self-assembly of lipid-DNA or polymer-DNA complexes in aqueous solution. X-ray studies have identified two key packing geometries: lamellar and hexagonal (May and Ben-Shaul 2004). In a lamellar structure, DNA double-stranded (dsDNA) molecules are sandwiched between lipid bilayers. In a hexagonal structure, the dsDNA molecules are intercalated within a hexagonal or honeycomb-like matrix. These structures may give rise to higher-order structures in solution. Lipid phase behavior has been reviewed more thoroughly elsewhere (May and Ben-Shaul 2004; Ewert et al. 2004). Complex stability is highly dependent on many factors: the lipid structure (Ren et al. 2000; Mintzer and Simanek 2009), the ratio of vector to DNA (Thierry et al. 1997), the structure of the DNA itself (length, circular, or linear), and environmental factors such as temperature, pH, and salt concentration. The structure of the lipid composition heavily influences transfection efficiency as it determines the degree of interaction with the cellular membrane and DNA cargo release. Neutral helper lipids such as dioleoylphosphatidylethanolamine (DOPE) (Boussif et al. 1995) can be added to improve transfection. Enhanced transfection efficiency due to DOPE may be due to its destabilization of lipid membranes, first proposed in 1995 (Farhood et al. 1995). DOPE and similar helper lipids form inverted hexagonal structural domains when complexed with DNA, which is linked with its higher transfection efficiency (Koltover et al. 1998). Investigations (Koltover et al. 1998; Hui et al. 1996; Lin et al. 2003; Mochizuki et al. 2013) suggest that the presence of DOPE or similar helper lipids may increase fusion between the lipid complex and the endosomal compartment thereby facilitating endosomal escape and DNA release.

Two landmark cationic lipids, *N*-[1-(2,3-dioleyloxy)propyl]-*N,N,N*-trimethylammonium chloride (DOTMA) (Felgner et al. 1987) and *N*-[1-(2,3-dioleoyloxy)propyl]-*N,N,N*-trimethylammonium methyl sulfate (DOTAP) (Stamatatos et al. 1988), have greatly shaped the development of lipofection. DOTMA, DOTAP and their derivatives are now widely commercially available. They share an overall structure: a quaternary ammonium head group and two hydrocarbon tails linked to a glycerol backbone. Structural modifications have been shown to have direct impacts on transfection efficiency; notably, close positioning of the cationic head group to the hydrocarbon chains were observed to be necessary for higher transfection activity (Ren et al. 2000). Another cationic lipid, 2,3-dioleyloxy-*N*-[2-(sperminecarboxamido)ethyl]-*N,N*-dimethyl-1-propanaminium trifluoroacetate (DOSPA) (Gebeyehu et al. 1992) became the basis for another widely used commercial transfection reagent, Lipofectamine.

Cationic polymers encompass another large non-viral vector space: positively charged polypeptides capable of DNA complexation and cell membrane interaction. Poly-L-lysine (PLL) was one of the first polymers identified for gene transfer applications (Laemmli 1975). Comprised of long lysine peptide chains, PLL has been investigated in both linear and branched forms (Okuda et al. 2004; Yamagata et al. 2007). Hyperbranched PLL may further provide greater transfection efficiency (Kadlecova et al. 2013) as greater molecular weight has correlated with higher transfection efficiency, though also greater toxicity. As PLL is highly positively charged, PEGylation is useful in reducing its toxicity and increasing stability (Männistö et al. 2002). Another cationic polymer, polyethylenimine (PEI) has been extensively studied since its introduction as a transfection agent in 1995 (Boussif et al. 1995) and is now considered the gold standard for transfection. Available as both a linear and branched polymer, it has a high buffering capacity, which contributes to its ability to escape from the endosomal pathway after cell uptake (Behr 1997).

Many more examples of cationic lipids and polymers have been reviewed in depth elsewhere (Mintzer and Simanek 2009). More recently, hybrid vectors comprised of both polymer and lipid agents (lipopolyplexes) have been investigated (Rezaee et al. 2016). Modified and unmodified PEI has been complexed with DOTMA (Matsumoto), DOTAP (Lee), and DOPE. The potential synergistic interactions between the polymer and lipid agent are being elucidated. Hybrid viral/lipid and viral/polymer vectors (Keswani et al. 2015) may be able to combine the best of both worlds: the highly efficient targeting and cell-penetrating abilities of viral peptides with the safety and low immunogenicity of the non-viral components.

Transfection into cell culture and in vivo using cationic lipids and polymers has generally been impaired by their high toxicity. Due to their highly charged surfaces, cationic lipid and cationic polymeric particles are rapidly cleared by the mononuclear phagocyte system (MPS) or are subject to aggregation by serum components (Senior et al. 1991; Chonn et al. 1992; Zelphati et al. 1998). To combat this, polyethylene glycol (PEG), an amphiphilic molecule, can be conjugated to the nanoparticle in both linear and branched forms. PEG [reviewed in Suk et al. (2016)] has high biocompatibility, low toxicity, and low immunogenicity and is very easy to manipulate, making it an ideal molecule for nanoparticle conjugation (Winkler 2015). Modification with PEG chains, also known as PEGylation, has been shown to help protect against clearance and serum aggregation (Harasym et al. 1995; Meyer et al. 1998; Rejman et al. 2004). This has led to the rise of PEGylated liposomes or "stealth liposomes," long-circulating liposomes capable of evading the immune system, which are better reviewed by Allen (1994) and more recently by Immordino et al. (2006). PEGylation has also seen much benefit in prolonging the circulation duration of highly cationic polyplexes as well, such as PLL and PEI (Ogris et al. 1999; Mishra et al. 2004). While highly beneficial for extending the serum lifetime of liposomal or polymer vectors, it has been noted that PEGylation may decrease transfection efficiency as it may impair uptake of the vector into targeted cells.

Uptake of most non-viral gene delivery vectors is known to be primarily by endocytosis (Zabner et al. 1995). After charge interaction between the cell surface and the nanoparticle, the particle is engulfed by one of many endocytic pathways. Liposomes have been visualized within the endosomal compartment by electron microscopy (Zhou and Huang 1994; Zabner et al. 1995) and confocal microscopy (Zabner et al. 1995). Pinocytosis and phagocytosis are thought to play a minor role in lipoplex uptake by specialized cells (Matsui et al. 1997). The major bottlenecks to chemical vector-mediated gene transfer once the vector breaches the cell membrane are endosomal escape and bypassing the nuclear membrane. Endosomal escape may be mediated through fusion or penetration of the endosomal membrane using peptides derived from viruses known to escape from the endosome. Such peptides include hemagglutinin (HAA) derived from influenza, TAT derived from HIV, and many others. However, the exposure of such antigenic peptides to serum and how to link these peptides to the vector must then be considered. Alternatively, some lipid and polymer agents with high buffering capacity, most notably PEI, have exhibited effective endosomal escape through the "proton sponge effect" (Behr 1997). Low pH within the endosome leads to increased protonation of the captured lipoplex or polyplex. This is thought to lead to an excessive inflow of ions and water into the endosome causing it to swell and rupture, releasing its contents into the cytoplasm.

The nuclear membrane presents another barrier to transfection in nondividing cells (van der Aa et al. 2006; Lam and Dean 2010). Non-viral gene delivery and expression is far more successful during mitosis when the nuclear membrane has broken down (Tseng et al. 1997). In quiescent cells, transport across the nuclear membrane is regulated by the nuclear pore complex (NPC) (van der Aa et al. 2006). These NPCs shuttle macromolecules (>40 kDa) between the cytoplasm and the nuclear compartment via specific nuclear import and export proteins. Import proteins recognize nuclear localization signals (NLS), typically short clusters of basic residues. A number of both peptide and DNA NLS sequences have been discovered (Escriou et al. 2003; Miller and Dean 2009). The combination of translocation sequences to overcome intracellular barriers with the appropriate cationic carrier to penetrate cellular barriers will be of great use to non-viral gene delivery.

Overall, cationic lipids and polymers comprise the largest category of non-viral gene delivery vectors, with many rapid innovations within the development field. Successful vectors such as PEI have shown great capability of both cell uptake and endosomal escape, while many other polymers and lipids have been far less efficient at navigating cellular barriers. Future innovations will therefore highly depend on elucidating the mechanisms behind these successes and also in improving the ability of non-viral vectors to bypass intercellular barriers.

2.4 Magnetofection

Magnetofection describes nucleic acid delivery mediated by the application of magnetic fields to susceptible vectors (Plank et al. 2000, 2011; Scherer and Plank 2008; Scherer et al. 2002; Mykhaylyk et al. 2007). Although development of magnetically assisted gene delivery stemmed independently by multiple groups after advancements made in magnetic drug targeting as early as the late 1970s (Mykhaylyk et al. 2007), the term "magnetofection" was first coined in by Plank in 2000 (Plank et al. 2000). Early successes paved the way for a plethora of developments both in vitro and in vivo, involving magnetofection; this contributes to the development of numerous commercially available products. As such, magnetofection poses to be a vital research tool in many applications of non-viral gene delivery.

Magnetofection involves the addition of magnetic vectors to cells/tissues followed by the application of permanent (static) or pulsating/oscillating (dynamic) magnets for magnetic field induction (Fig. 2.1). High transfection efficiency associated with magnetofection is mainly attributed to the rapid sedimentation of magnetic vectors onto cell surfaces upon implementation of a magnetic field (Mykhaylyk et al. 2007; Plank et al. 2011; Sauer et al. 2009; Kami et al. 2011; Steitz et al. 2007; Liu et al. 2011a; Song et al. 2010; McBain et al. 2008; Vainauska et al. 2012; Sun et al. 2012; Kamau et al. 2006). The application of magnetic fields improves transfection kinetics by reducing time for transfection from hours to minutes. Rapid sedimentation serves to overcome the diffusion barrier that limits vector accumulation at target cells and, consequently, lowers the required vector concentrations at target tissues (Scherer et al. 2002; Plank et al. 2011; Vainauska et al. 2012). Thus, magnetofection offers several significant advantages: (1) rapid vector internalization, (2) high transfection efficiencies at lower vector doses, (3) lower cytotoxicity, (4) reduced incubation time, and (5) the potential for synchronized transfections. After improved accumulation at the cell membrane, cell uptake of the non-viral vector would then proceed through endocytic routes as is typical for the vector in question (Plank et al. 2011). Cai et al. (2005) examined magnetic carbon nanotubes to mechanically penetrate past cell membranes using a physical means of entry. This will be further examined in the carbon nanomaterials section. Rational vector design remains

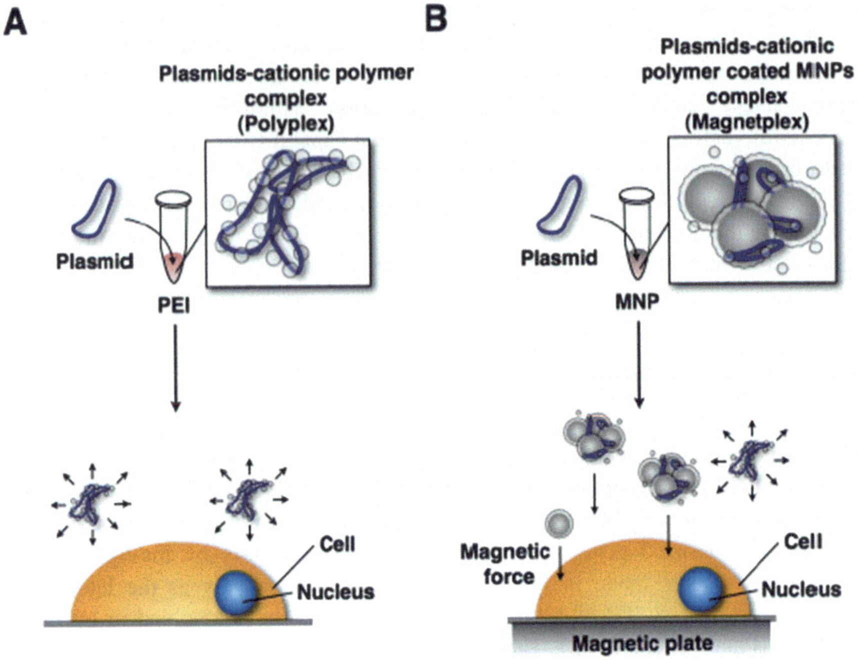

Fig. 2.1 (**a**) Standard transfection process mediated by plasmids-cationic polymer complex (polyplex); (**b**) transfection process mediated by plasmids-cationic polymer-coated MNPs complex (magnetoplex). Adapted from Kami et al. (2011)

crucial in attaining high transfection efficiencies. Since non-viral magnetic vectors are assembled in a combinatorial manner, the different components (magnetic core, nanoparticle coating, transfection vector, and nucleic acid cargo) can be individually optimized to tailor to different specific applications (Scherer et al. 2002).

For magnetic nanoparticles (MNPs) to be deemed applicable for magnetofection, they must possess four key characteristics: (1) the ability to interact with nucleic acid or with nucleic acid vector, (2) sufficient magnetic properties to drive rapid sedimentation by affordable and relatively mobile magnetic devices, (3) high stability for long-term storage, and (4) biocompatibility for application in living cells/tissues (Mykhaylyk et al. 2007). Typically, superparamagnetic iron oxide nanoparticles (SPIONs) are employed as the magnetic core due to their superparamagnetic properties, high biocompatibility, low toxicity, biodegradability in vivo, and concurrent applications in magnetic resonance imaging (MRI) (Plank et al. 2011; Kami et al. 2011). SPIONs are typically coated by different materials including carbohydrates (dextran, chitosan, heparin sulfate), synthetic polymers [polyethylenimine (PEI), polyethylene glycol (PEG), dendrimers, poly(lactic-co-glycolic acid) (PLGA)], proteins (albumin, streptavidin), cationic cell-penetrating peptides (e.g., TAT peptide), surfactants, phospholipids, silica, and gold (Kami et al. 2011; Wahajuddin 2012). The surface coating reduces MNP aggregation, prevents oxidation, lowers

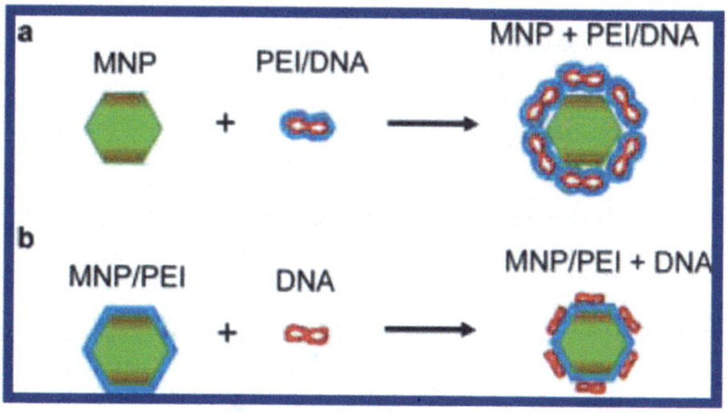

Fig. 2.2 Proposed structures of fluorescent MNP-PEI-DNA hybrid structures. Adapted from Arsianti et al. (2010)

cytotoxicity, and enables functionalization of MNPs. SPIONs are commonly coated with PEI, well known as a gene carrier with high transfection efficiency but also associated with high cytotoxicity. PEI-coated MNPs are commonly used in transfection studies as evidenced by several commercially available products (transMAG ^{PEI}, PolyMAG, CombiMAG). Lipid-coated MNPs, or magnetoliposomes, can be generated by encasing the magnetic core with a cationic shell composed of various mixtures of cationic and neutral lipids/phospholipids (Yang et al. 2008; Namiki et al. 2009).

The formation of non-viral magnetic vectors involves the association of MNPs with lipid- or polymer-based transfection vector, predominantly through electrostatic interaction (Mykhaylyk et al. 2007). As MNPs are commonly coated with materials pertinent to transfection (e.g., MNP^{PEI} and cationic lipids), magnetic vectors can be assembled through direct interaction between coated MNPs and nucleic acid. By themselves, MNP/PEI/DNA or MNP^{PEI}/DNA magnetoplexes and magnetoliposomes can self-assemble into effective magnetic vectors (Fig. 2.2); however, additional components are usually incorporated to assist with the transfection process. The incorporation of PEG is advantageous in in vivo applications as it reduces serum protein interactions, thereby prolonging vector half-life; however, addition of PEG has been shown to dampen gene delivery possibly due to interference with intracellular trafficking (Cool et al. 2013). The incorporation of TAT peptides permits efficient intracellular trafficking resulting in further enhancements to transfection efficiencies. Subsequent fusion of histidine residues to TAT peptide permits effective endosomal escape through the proton sponge effect as histidine residues adsorb protons within the acidic environments of endolysosomes. Using histidine-rich TAT peptides, Song et al. (2010) demonstrate in vitro gene delivery of self-assembled MNP^{PEI}(PolyMAG)/DNA/TAT (histidine-rich) ternary complexes into human astrocytoma cells, yielding transfection efficiencies as high as 60% at 12:1 TAT/DNA ratio. Subsequent in vivo delivery of the ternary complexes into a rat spinal cord by lumbar injection resulted in a twofold increase in transgene expression when compared to MNP^{PEI}(PolyMAG)/DNA (Song et al. 2010). As cellular uptake of magnetic vectors is predominantly through endocytosis (Scherer et al. 2002; Plank et al. 2011; Sauer et al. 2009), functionalizing magnetic vectors with ligands allows for targeted delivery into specific cells. These ligands can be

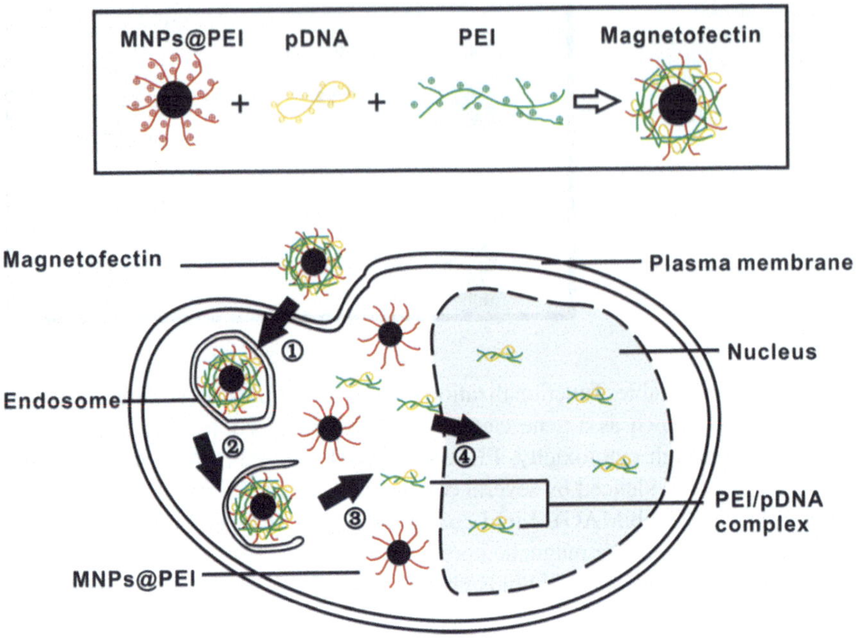

Fig. 2.3 Intracellular events of MNP^{PEI}/PEI/plasmid DNA (pDNA) magnetofection: (1) cellular uptake; (2) endosomal escape; (3) intracellular trafficking highlighting the dissociation of MNP^{PEI} from PEI/pDNA; and (4) nuclear entry by PEI/pDNA. Adapted from Ma et al. (2011)

incorporated in several different methods: (1) electrostatic interactions, (2) conjugation to a preexisting component to be incorporated into the vector (e.g., PEG), or (3) direct conjugation to MNPs by a biotin-streptavidin bridge. Surface characteristics of magnetic vectors heavily influence serum protein interactions, aggregation behavior, cellular uptake, and intracellular trafficking (Fig. 2.3) (Arsianti et al. 2010). Since cellular uptake of magnetic vectors is mediated by endocytosis, the size of the vector plays an important factor in transfection efficiency as it affects the extent of uptake and intracellular trafficking (Plank et al. 2011; Steitz et al. 2007; Park et al. 2011). Both size and surface characteristics must be taken into consideration in order to achieve high transfection efficiencies. Therefore, controlling the ratios of each magnetic vector component must be optimized.

Since magnetic vectors are assembled in a combinatorial manner, formation of lipoplexes and polyplexes can precede the addition of MNPs during vector assembly. By mixing the order of the different vector components, the vector's biophysical properties are affected, thus impacting transfection efficiency. With respect to PEI, DNA, and MNP^{PEI} compositions, vectors with surface PEI (positive zeta potential) achieve effective gene delivery through rapid cellular uptake and release from endolysosomes; on the other hand, vectors with surface DNA (negative zeta potential) have inefficient gene delivery due to inadequate cellular uptake and endolysosomal release (Arsianti et al. 2010). Formation of lipoplex/polyplex prior to MNP addition may ensure optimal surface characteristics for better transfection efficiency. A study by Ma et al. (2011) indicates that only free PEI/pDNA polyplexes may enter the nucleus, while MNP^{PEI} remains within the cytoplasm. In addition, transfection of SPC-A1 and COS-7 cells is hindered by the absence of free PEI/pDNA polyplexes; this implicates the requirement of PEI/DNA polyplexes to dissociate from MNP for nuclear entry (Ma et al. 2011). Therefore, it may be beneficial to form lipoplexes/polyplexes prior to MNP addition during vector assembly. Overall, component selection and assembly method are both critical to the design of a magnetic vector.

Early literature results suggested that magnetofection has no influence on the uptake and subsequent intracellular trafficking of magnetic vectors. Further analysis by Sauer et al. (2009) confirmed this with single-particle analysis; assessing the intracellular fate of magnetic lipoplexes indicated the same three distinct phases present in both magnetic and nonmagnetic PEI/DNA polyplexes. Upon rapid sedimentation, Sauer et al. observed (1) cellular uptake by endocytosis, (2) confined diffusion in cytoplasm, and (3) microtubule-mediated active transport of vesicle-bound magnetic lipoplexes. The observation of a perinuclear ring confirms the intracellular trafficking of the vesicle-bound magnetic lipoplexes into the perinuclear region. However, endosomal escape and MNP dissociation remained problematic as the majority of magnetic lipoplexes remained vesicle-bound after 24 h with 90% of the complexes showing no separation between pDNA and MNPs (Sauer et al. 2009). In contrast, Xie et al. (2015) more recently examined transfection of HepG2 cells with magnetic PEI/DNA complexes (MPD-cc) using fluorescent imaging. Although transfection efficiency was overall low (11.3%), transfection with application of a magnetic field was far superior than without (0.7%). More importantly, visualization of Cy5-labeled DNA and FITC-labeled PEI with confocal microscopy demonstrated destruction of the endolysosomal compartment after 24 h, indicating improvement in endosome escape using magnetofection. Xie et al. concluded the increased transfection efficiency could be attributed to lower protein adsorption when within a magnetic field (Xie et al. 2015). These results in contrast to previous results indicate that more factors could be in play, including target cell type, cargo DNA type, degree of magnetic field, and the composition of the magnetic complexes.

Magnetic particle size, composition and concentration, magnetic field strength, and exposure times are key parameters that need to be optimized for enhanced transfection efficiency (Stride et al. 2009). Magnetofection generally involves the application of a permanent (static) magnetic field with a field strength ranging from 70 to 250 mT and a field gradient ranging from 50 to 130 T m^{-1} (Mykhaylyk et al. 2007). It has been proposed that pulsating/oscillating (dynamic) magnetic fields introduce lateral, rolling motion of the magnetic vectors upon rapid sedimentation, and such motions may prevent large aggregate formation, enable adequate distribution of sedimenting magnetic vectors, induce receptor clustering, and apply mechanical stimuli facilitating endocytosis (McBain et al. 2008; Vainauska et al. 2012; Kamau et al. 2006). At optimal conditions, the application of pulsating/oscillating magnetic fields, alone or in combination with permanent magnetic fields, has shown

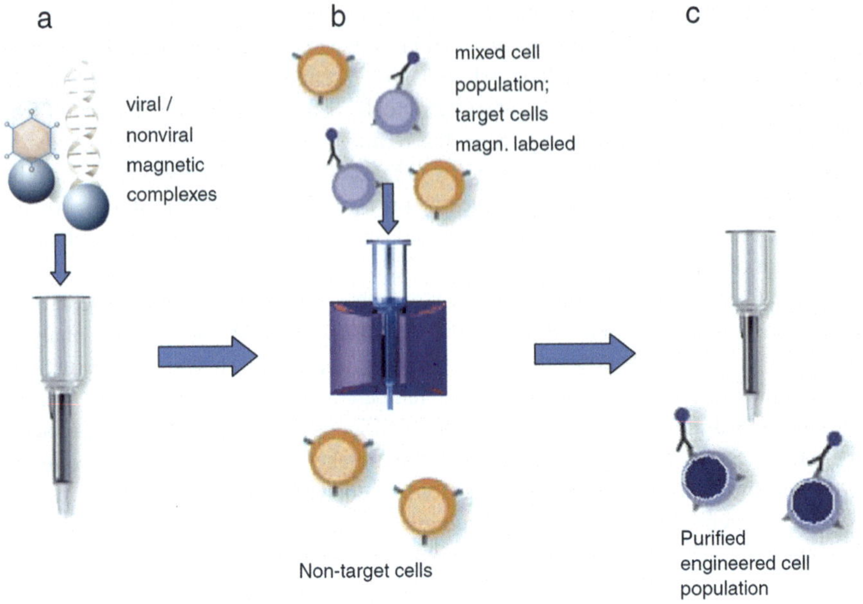

Fig. 2.4 "Magselectofection." Adapted from Al-Dosari and Gao (2009)

to improve transfection efficiencies when compared to magnetofection solely by permanent magnetic fields (Steitz et al. 2007; McBain et al. 2008; Vainauska et al. 2012; Kamau et al. 2006).

Alternatively, magnetofection has been conducted in conjunction with magnetic cell sorting through a process referred to as "magselectofection" (Fig. 2.4) (Plank et al. 2011). A high-gradient magnetic field cell separation column is used as the magnetic device for simultaneous cell sorting and magnetofection. Magnetic vectors and a mixed cell population, composed of both magnetically labeled and unlabeled cells, are added and passed through the separation column. The induced magnetic fields promote close interactions between magnetic vectors and magnetically labeled cells which results in high transfection efficiencies (Plank et al. 2011). Cell separation is achieved as magnetically labeled cells remain within the column while non-labeled cells pass through the column.

Studies have indicated that, in the presence of a magnetic field, MNP/PEI/DNA- or MNP^{PEI}/DNA-based magnetoplexes can induce rapid internalization and enhance transfection, yielding efficiencies greater than transfections mediated by Lipofectamine (Song et al. 2010; Vainauska et al. 2012; Miao et al. 2013; Al-Deen et al. 2011; Ensenauer et al. 2011; Kong et al. 2012a; Wang et al. 2011) and PEI (Park et al. 2011; Miao et al. 2013; Namgung et al. 2010; Prijic et al. 2012) while exhibiting lower cytotoxicity than PEI (Steitz et al. 2007; Song et al. 2010; Sun et al. 2012; Park et al. 2011; Al-Deen et al. 2011; Namgung et al. 2010). As expected, transfection efficiency is also shown to be enhanced by the implementation of magnetoliposomes over traditional liposomes through magnetofection (Yang et al. 2008; Prijic et al. 2012). For traditional transfection agents (e.g., PEI and cationic lipids), higher transfection efficiencies are generally associated with greater cytotoxicity from increasing concentrations. Higher concentrations are required such that a sufficient N/P ratio (molar ratio of the complexing cationic polymers to the anionic phosphate groups of the nucleic acid) could be attained for effective compaction and encapsulation of nucleic acid. With respect to MNP-PEI-DNA-based magnetoplexes, and magnetic vectors in general, lower cytotoxicity is mainly attributed to effective internalization induced by rapid sedimentation. Through magnetofection, lower nucleic acid dose is required for effective transfection efficiencies; this corresponds to lower concentrations of the transfection agent and lower cytotoxicity. The benefits of magnetofection are clearly highlighted by enhanced transfection efficiencies in the presence of serum proteins as rapid sedimentation limited nonspecific interactions between magnetic vectors and serum proteins (Liu et al. 2011a; Sun et al. 2012; Al-Deen et al. 2011; Ensenauer et al. 2011). In the presence of serum proteins, interactions between negatively charged serum protein and surface PEI induce surface charge reversal and disruption of the magnetic vector (Arsianti et al. 2010). Nonspecific serum protein interactions present an important roadblock for successful translation of in vitro findings to in vivo settings. This imposes significant implications to in vivo applications of magnetic vectors.

Magnetofections have been employed in various different applications of gene delivery in vitro including transfection of primary cells (epithelial, endothelial, cardiomyocytes, myoblasts, chondrocytes, fibroblast, synoviocytes, osteoblasts, melanocytes, and mouse embryonic stem cells) (Plank et al. 2011), transfection of neural precursor/stem cells (Sapet et al. 2011; Jenkins et al. 2011; Pickard et al. 2011; Fallini et al. 2010), and induced pluripotent stem cell generation (Park et al. 2012). Magnetofections have been applied to in vivo applications, particularly in cancer gene therapy due to tumors' leaky vasculature permitting both passive targeting, by enhanced permeability and retention effect, and active targeting by magnetic force and ligand-receptor interactions (Plank et al. 2011). Magnetofection studies involving the delivery of plasmids coding for tumor necrosis factor-related apoptosis ligand (TRAIL) (Miao et al. 2013) and murine interleukin 12 (IL-12) (Prijic et al. 2012) both report significant tumor reductions and antitumor effect in mice. Similarly, tumor growth rate in mice has been significantly lowered as a result of downregulated expression of type 1 insulin-like growth factor receptor (IGF-1R) through magnetofection of IGF-1R siRNA (Kong et al. 2012a). Through magnetofection, Zhou et al. (2007) were able to demonstrate enhanced immune response to HIV-DNA vaccine in mice through the delivery of HIV-gag encoding magnetic vectors by tibialis anterior injections and subsequent magnetofection. Using magnetofection, immune response is activated at low doses corresponding to 1000-fold increase in DNA vaccine potency. A single immunization by magnetofection was able to elicit 100% seroconversion after 1 week in contrast to 5 weeks after three separate immunizations of naked DNA; this highlighted the ability of magnetofection to accelerate humoral immune response. Subsequent in vitro analysis confirms the potential of cytotoxic T-lymphocyte response upon immunization (Zhou et al. 2007). In addition to enhancing transfection efficiency by magnetofection, the application of an external magnetic field allows for localized and guided targeting in regions specifically exposed to the magnetic field. This feature was highlighted by Song et al. (2010) where moving the external magnetic field down the spinal cord changed the regional distribution pattern of transgene expression.

Magnetofection presents a highly promising avenue of increased vector accumulation to targeted cells. However, despite the advantages of magnetofection in vitro and in vivo, the technique is still limited by the low penetration depth of the external magnetic field, high field strengths required to overcome hydrodynamic forces for MNP localization, and insufficient biodistribution analysis demonstrating the effects of magnetic targeting (Plank et al. 2011; Owen et al. 2012). In order to truly realize the benefits of in vivo magnetofection, these obstacles must be overcome for deeper tissue penetration.

2.5 Sonoporation: Nano-/Microbubbles and Bubble Liposomes

Ultrasound consists of acoustic waves that propagate at frequencies beyond the human auditory range. For medical applications, ultrasound has been commonly used for ultrasonography, a noninvasive diagnostic medical imaging technique capable of visualizing internal structures and organs in real time. In comparison to other imaging techniques—computed tomography (CT), positron emission tomography (PET), and magnetic resonance imaging (MRI)—ultrasound is advantageous in terms of its low costs, portability, ease of use, and suitability for repetitive administrations in the absence of ionizing radiation. The low power levels involved in ultrasound imaging limit the risk of any significant adverse effects on tissues. Conventional ultrasound imaging employs sound wave pulses, transmitted from a probe (transducer) at a specific frequency. These sound waves become reflected and scattered as they pass through tissues. Scattering is caused by variations in density and compressibility of the different tissue components; this results in contrast and the generation of an ultrasound image (Alzaraa et al. 2012). Images can be refined by the addition of contrast agents that serve to increase scattering and reflection of ultrasound signals (Böhmer et al. 2009).

In addition to diagnostic imaging, ultrasound can be applied in therapeutics through thermal and nonthermal (pressure) effects associated with sound waves. Thermal effects can be induced upon the application of high-intensity focused ultrasound (HIFU) for the absorption of acoustic energy by fluids and tissues, resulting in an increase in localized temperature and tissue damage (Alzaraa et al. 2012; Nomikou and McHale 2012; Husseini and Pitt 2008; Feril and Tachibana 2012). Thermal effects associated with HIFU are the basis of hyperthermic ablation therapy, a type of cancer therapeutic administered in conjunction with systemic chemotherapy and radiotherapy (Alzaraa et al. 2012; Nomikou and McHale 2012;

Sanches et al. 2011). Temporal induction of localized hyperthermia has also been used in drug/gene delivery by means of temperature controlled release and subsequent induction of elevated transgene expression (Böhmer et al. 2009; Feril and Tachibana 2012). Nonthermal effects of ultrasound are associated with the acoustic pressure waves generated from low-intensity ultrasound, and it is the basis of thrombolysis therapy (Kiessling et al. 2012) and sonoporation.

Sonoporation is a physical method of non-viral gene delivery through modifications to the endothelial layer and the cell membrane permeability as a result of the application of low-intensity ultrasound (Fig. 2.7); thus, sonoporation facilitates gene delivery by enhancing extravasation and internalization. Transient pore formation and upregulation of endocytosis are speculated to be the key mechanisms of gene delivery by sonoporation (Delalande et al. 2012). Transient pore formation is proposed to occur by several different ways: (1) mechanical stimulation (pushing and pulling effects on cell membrane), (2) shear stress and microstreams, (3) microjets/jetting, (4) sonophore formation, and (5) microbubble translation (Delalande et al. 2012). Pore formation may permit direct entry of nucleic acids into the cytoplasm and has been postulated to cause transient intracellular Ca^{+2} influxes that result in upregulation of endocytosis (Delalande et al. 2012). Other associated effects resulting from transient pore formation include increases in membrane permeability by mechanical stimulation and reactive oxygen species as well as increases in membrane fluidity by minor elevations in temperature (Alzaraa et al. 2012; Husseini and Pitt 2008; Feril and Tachibana 2012; Kiessling et al. 2012; Delalande et al. 2012; Yoon and Park 2010).

At very low ultrasound intensities, the acoustic pressures generated by ultrasound waves induce certain membrane effects that may lead to an increase in permeability and pore formation. The bilayer sonophore model, introduced by Krasovitski et al. (2011), suggests that cell membranes are capable of converting acoustic pressure wave into mechanical forces that act to cause expansions and compressions of the intramembrane space; transient pore formation can result from the growing tension on the membrane leaflets. Sonophores have been shown to arise from membranes of subcellular compartments (Krasovitski et al. 2011). However, sonophores alone may not be able to cause sufficient membrane permeability/disruption required for effective gene delivery. The addition of microbubbles, contrast agents used in ultrasound imaging, effectively introduces and amplifies linear/nonlinear oscillations for effective ultrasound-mediated gene delivery. Microbubbles have been observed to undergo oscillations causing local deformation and transient pore formation at low ultrasound intensities (Moosavi Nejad et al. 2011). Through observations with fluorescently labeling microbubbles, Delalande et al. (2011) proposed a new mechanism involving the direct translation of microbubbles across cell membranes; microbubbles have been shown to enter cells and rapidly dissolve, losing a portion of its shell during the process.

The extent of microbubble oscillations is under the influence of applied acoustic pressures, which is proportional to ultrasound intensity (Fig. 2.5) (Nomikou and McHale 2012). At higher acoustic pressures, microbubbles introduce and amplify nonlinear oscillations that give rise to cavitation. Depending on the ultrasound

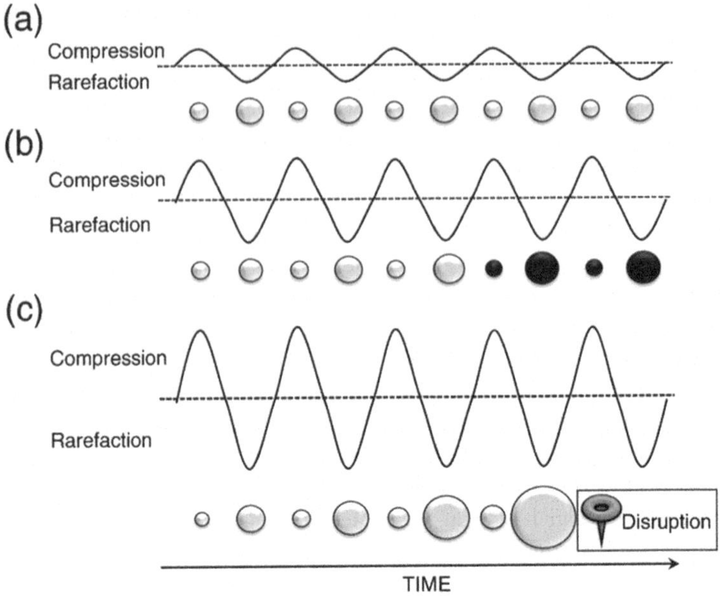

Fig. 2.5 Microbubbles in response to increasing ultrasound intensities: (**a**) very low-intensity ultrasound induces linear oscillations; (**b**) low-intensity ultrasound induces nonlinear oscillations and microbubble growth; stable oscillations occur at resonant size; (**c**) high-intensity ultrasound induces substantial nonlinear oscillations and for bubbles to collapse. Adapted from Suzuki et al. (2011)

intensity, cavitation is subdivided into two categories: stable and inertial cavitation (Fig. 2.6). Stable cavitation occurs at intensities where continuous compression and expansion of microbubbles give rise to nonlinear oscillations leading to the formation of microstreams. Microstreams in the surrounding region induce shear stress that creates transient pores, increases membrane permeability, and upregulates pathways associated with stress response (Cool et al. 2013; Nomikou and McHale 2012; Husseini and Pitt 2008; Delalande et al. 2012; Suzuki et al. 2011; Qiu et al. 2012). The microbubbles at this point remain stable and grow in size by accumulating dissolved gases. They continue to grow until they reach resonant size, resulting in stable oscillations (Husseini and Pitt 2008; Suzuki et al. 2011). At higher intensities, forced expansion and compression of microbubbles cause bubbles to collapse and inertial cavitation. Inertial cavitation creates substantial nonlinear oscillations and leads to the formation of shockwaves and microjets that give rise to transient pores on cell membranes (Cool et al. 2013; Nomikou and McHale 2012; Husseini and Pitt 2008; Suzuki et al. 2011; Qiu et al. 2012). Cell membrane deformations caused by oscillating microbubbles has been previously captured by ultrafast camera; thus, the process of transient pore formation by stable and inertial cavitation has been demonstrated (Böhmer et al. 2009). Studies by Qiu et al. (2010, 2012) have indicated the influence of inertial cavitation in sonoporation from

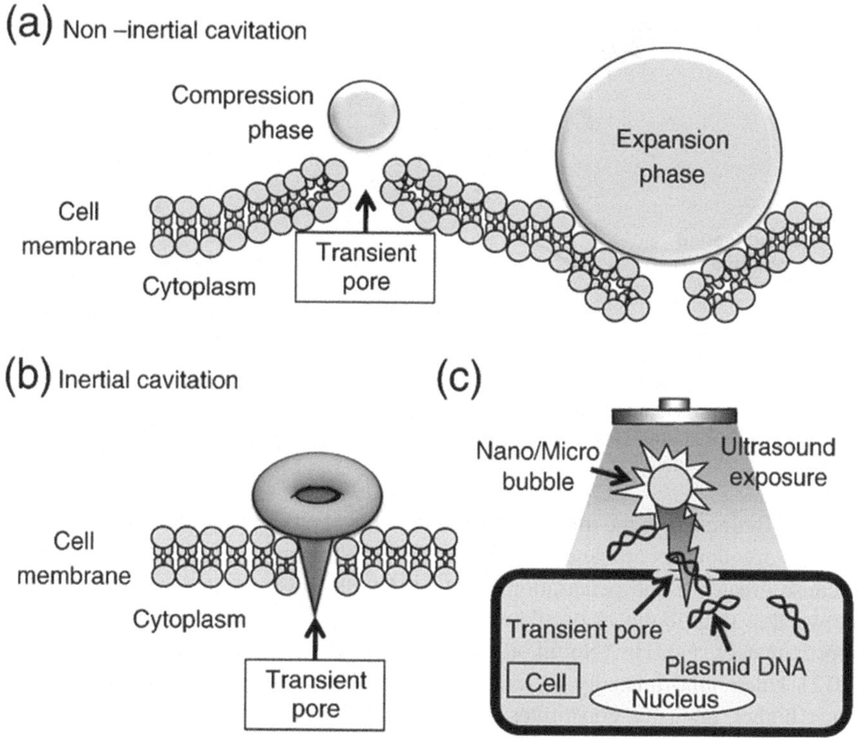

Fig. 2.6 The effects of (**a**) stable and (**b**) inertial cavitation on cell membranes. Adapted from Suzuki et al. (2011)

evidence indicating that there is a high correlation between inertial cavitation, transfection efficiency, and cell toxicity. Initially, inertial cavitation is linearly correlated with transfection efficiency but becomes saturated at a certain point; this is an indication of excessive inertial cavitation leading to permanent membrane damage and cell death. Increasing acoustic pressure or duration has been correlated to increases in pore size and higher transfection at a cost of increasing cytotoxicity (Qiu et al. 2012). Hence, ultrasound parameters need to be optimized such that sufficient ultrasound exposure is applied to enhance transfection efficiency without compromising cell viability.

Frequency, intensity, duration, microbubble composition, microbubble and nucleic acid concentrations, and cell type are key parameters associated with ultrasound-mediated gene delivery (Fig. 2.7) (Delalande et al. 2012; Yoon and Park 2010; Suzuki et al. 2011; Figueiredo and Esenaliev 2012). Generally, high ultrasound frequencies, ranging from 3 to 10 MHz, are used for clinical imaging, while lower ultrasound frequencies, ranging from 1 to 2 MHz, are used for conventional therapeutic ultrasound (Yoon and Park 2010). The most optimal frequency for gene delivery is dependent on optimal microbubble interactions and propagation path. The highest level of oscillation occurs when the applied frequency matches the

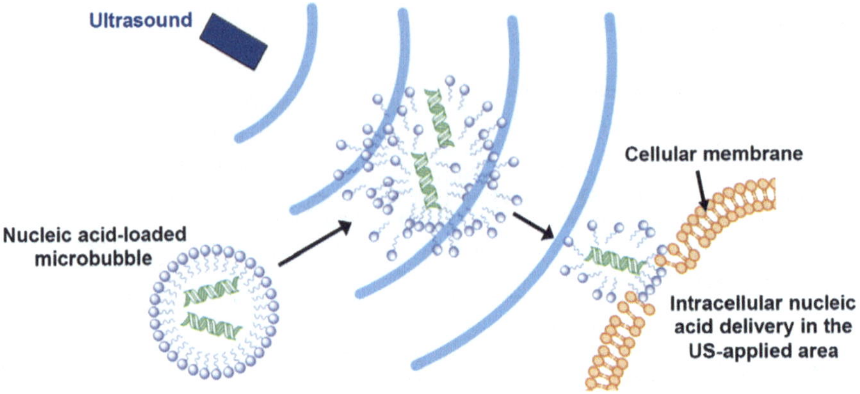

Fig. 2.7 Sonoporation-mediated gene delivery. Adapted from Shim and Kwon (2012)

resonant frequency of the microbubble (Böhmer et al. 2009; Husseini and Pitt 2008). The low frequencies are generally employed for ultrasound-mediated gene delivery because greater tissue penetration is attained with lower frequencies. It has been shown that microbubbles oscillate steadily, with maximized tissue penetration at approximately 1 MHz (Nomikou and McHale 2012; Figueiredo and Esenaliev 2012). Ultrasound intensity (W/cm^2) is proportional to acoustic pressure squared; since higher intensity constitutes higher energy deposition, high-intensity ultrasounds (>5 W/cm^2) are suitable to induce thermal effects, while low-intensity ultrasounds (0.125–3 W/cm^2) are suitable to induce nonthermal effects (sonoporation) (Yoon and Park 2010). The mechanical index (MI), a measure of the likelihood of inertial cavitation, is defined as $MI = \frac{P^-}{\sqrt{f\ (\mathrm{MHz})}}$ where P^- is the peak rare-fractional pressure and f is the frequency (Husseini and Pitt 2008; Yoon and Park 2010). The MI must be carefully noted as it provides an indication of microbubble activity upon ultrasound exposure: MI < 0.05–0.1 (linear oscillations), 0.1 < MI < 0.3 (stable cavitation), MI > 0.3–0.6 (inertial cavitation), and MI > 1 (detrimental biological effects) (Husseini and Pitt 2008; Suzuki et al. 2011). Correlations between studies are difficult due to the variations in each parameter. Numerous in vivo studies have been conducted using a wide range of values for each parameter: frequency (1–4 MHz), intensity (0.5–5 W/cm^2), pulse mode (10%—continuous wave), and stimulation time (10 s–30 min) (Delalande et al. 2012). Varying responses to microbubble composition significantly compound issues relating to cross-study comparisons, as a large number of different microbubble formulations have been presented in literature, each with its own inherent distinct properties. This is further compounded by variations in cell-type responses to different microbubbles and ultrasound conditions.

Microbubbles, with sizes ranging from 1 to 10 μm, can be classified into four different groups: coated microbubbles, multilayered microbubbles, phase-shift emulsion, and echogenic liposome. Coated microbubbles consist of a biodegradable shell coating, which in itself may be composed of surfactant, protein, biopolymer, galactose, and/or lipids (Alzaraa et al. 2012), encapsulating gases with air or gases. Coating composition and coating consistency determines the stability of the coated microbubble as well as its sensitivity to pressure changes induced by ultrasound exposure (Alzaraa et al. 2012; Yoon and Park 2010; Nomikou et al. 2012). First-generation coated microbubbles are composed of air encased with surfactant, galactose, or denatured albumin coating; they are ineffective due to their short half-life and high MI (Alzaraa et al. 2012). Second-generation coated microbubbles, composed of a stronger coating and hydrophobic heavy inert gases (perfluorocarbon, perfluoropropane, or sulfur hexafluoride), have greater stability (longer half-life) and can reach the main arteries by crossing the pulmonary vascular bed (Alzaraa et al. 2012; Suzuki et al. 2011). Optison (Molecular Biosystems Inc.), Sonovue (Bracco Diagnostics), and Definity (Bristol-Myers Squibb Medical Imaging) are three second-generation microbubbles approved by the FDA for human use. The aforementioned microbubbles are thin-shelled microbubbles (2–3 nm thick) that require hydrophobic fluorinated gases to prevent gas dissolution for better stability; on the other hand, polymer microbubbles are thick-shelled microbubbles (20–100 nm thick) that do not require hydrophobic fluorinated gas for stability but do require higher acoustic pressures for oscillation and cavitation (Böhmer et al. 2009; Sanches et al. 2011; Delalande et al. 2012). Multilayered microbubbles serve to increase the effective surface area of the microbubble for improved functionalization. Phase-shift emulsion consists of stabilized emulsions of volatile liquid droplets that become microbubbles upon injection and subsequent ultrasound exposure (Yoon and Park 2010). Echogenic liposomes are gas containing-liposomes composed of a phospholipid bilayer (Huang 2008).

Ultrasound-mediated gene delivery by sonoporation can be accomplished by co-administration of microbubbles and nucleic acid as separate entities (Nomikou and McHale 2012). As sonoporation induces transient pores that allow direct entry of nucleic acids into the cytoplasm, nucleic acids are not required to be packaged into a transfection vector. However, high nucleic acid concentrations are required to elicit therapeutic effect due to rapid degradation of naked nucleic acid (Cool et al. 2013). Hence, packaging nucleic acid into a transfection vector is still warranted for encapsulation, condensation, and protection. Alternatively, the attachment of nucleic acid to microbubble protects the nucleic acid from degradation, improves circulation profile, and increases delivery upon ultrasound exposure (Sirsi et al. 2012). Regardless, co-administration of microbubbles and nucleic acids/transfection vectors (lipoplex or polyplex) constitutes low in vivo transfection efficiency due to differing pharmacokinetic profiles of microbubbles and vectors that render in vivo distribution of both components difficult to control (Un et al. 2010). As a result of this, nucleic acid and/or transfection vectors are generally attached to microbubbles for transport to specific sites, followed by release and delivery upon microbubble cavitation and transient pore formation. There are numerous methods of incorporating nucleic acid cargo into microbubbles: (1) direct physical incorporation into microbubble shell during fabrication, (2) electrostatic interaction with cationic lipids incorporated into microbubble shell, (3) formation of a multilayered microbubble

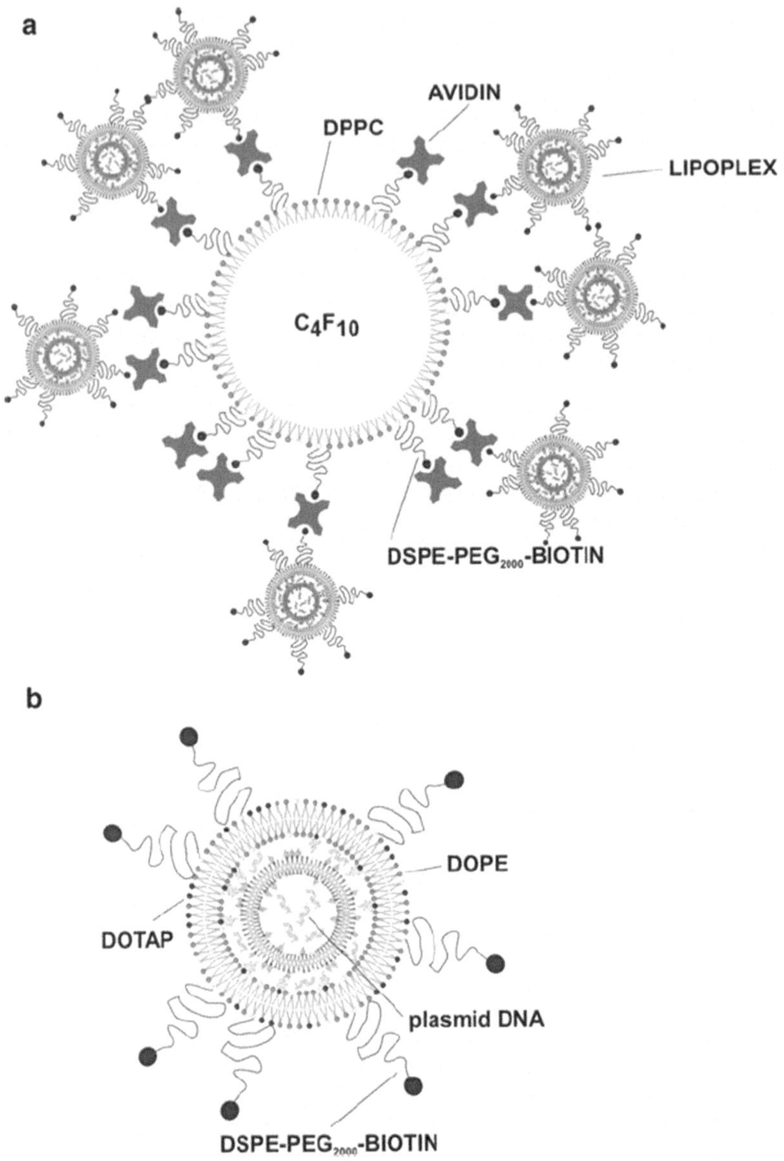

Fig. 2.8 PEGylated lipoplexes attached to microbubble surface by biotin-avidin linkage. Adapted from Cool et al. (2013)

with a surface layer of cationic polymer for nucleic acid binding, (4) tethering transfection vectors by hybridization of complementary DNA oligonucleotides, (5) covalently linking transfection vectors to microbubbles, and (6) linking transfection vectors to microbubbles through a strong non-covalent biotin-avidin bridge (Fig. 2.8) (Cool et al. 2013; Sirsi et al. 2012). Similarly, ligands can also be linked to microbubbles with biotin-avidin linkage and peptide bonds or become incorporated during microbubble synthesis (Böhmer et al. 2009; Kiessling et al. 2012). The addition of ligands permits active targeting to specific sites, which enhances site specificity in conjunction with ultrasound-mediated delivery. Thus, lower toxicities can be attained from the reduction in transfection vectors necessary for enhanced transfection efficiency.

Sonoporation has been employed for in vivo gene delivery into various cells/ tissues (endothelium, skeletal muscles, peritoneal macrophages, salivary gland, ocular ciliary muscle, myocardium, skin, carotid artery, and embryonic limb bud mesenchyme/ectoderm) (Nomikou and McHale 2012; Suzuki et al. 2011). siRNA delivery, limited by their inability to passively diffuse through cellular membranes, has been achieved through sonoporation in numerous documented studies (Cool et al. 2013; Böhmer et al. 2009; Yoon and Park 2010; Suzuki et al. 2011; Vandenbroucke et al. 2008). Similarly, shRNA has been documented to be delivered by sonoporation; this permits stable gene expression and prolong RNA interference (Chen et al. 2009a, b, 2010). In vivo studies in rabbits have demonstrated transient pore formation at the blood-brain barrier that persists for up to 6 h (Böhmer et al. 2009).

Focused ultrasound has been successfully employed in the noninvasive and reversible disruption of the blood-brain barrier (BBB), permitting drug or gene delivery to the brain (Timbie et al. 2015). MRI-guided ultrasound at low-pressure amplitude and low acoustic power has been shown to mediate transient pore formation at the BBB. Particles as large as 100 nm can then bypass the permeabilized BBB and circulate in the brain (Timbie et al. 2015). Nance et al. (2014) constructed stealth nanoparticles (PEGylated polystyrene particles) which they termed "brain-penetrating nanoparticles" (BPN) for delivery to the murine brain both ex vivo and in vivo. More recently, they demonstrated successful delivery of GDNF-expressing plasmid using BPN, eliciting robust and sustained gene expression over 10 weeks post treatment (Mead et al. 2017). These findings have serious implications on drug and gene delivery to the brain by focused ultrasound. Effective image-guided spatial and temporal control over ultrasound, by optimizing ultrasound parameters, enables site-specific gene delivery with negligible local and systemic toxicities (Sanches et al. 2011; Yoon and Park 2010; Sirsi et al. 2012). Hence, ultrasound is capable of increasing efficacy of non-viral vectors through targeted delivery into specific tissues. Repeat applications can occur without compromising cell/tissue viability since sonoporation is a transient process. Minimal invasiveness of ultrasound, high targeting capacity, and suitability for repeat administrations all contribute to numerous clinical applications as highlighted in reviews by Delalande et al. (2012), Geis et al. (2012) and Farhood et al. (1995).

Despite the fact that good acoustic responses can be attained with microbubbles, in vivo transfection efficiencies remain suboptimal due to their size limitations. The large size of microbubbles may cause them to be subjected to shear forces encountered in the bloodstream (Böhmer et al. 2009). Through stable cavitation, microbubbles can enhance extravasation; however, they maintain a relatively short half-life (20 min in humans; few minutes in mice or rats) (Böhmer et al. 2009) and

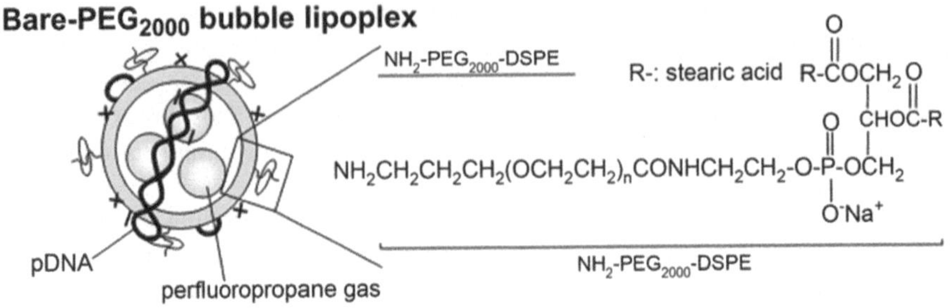

Fig. 2.9 Schematic of a bubble liposome. Adapted from Un et al. (2010)

are hampered by limited perfusion and gene delivery beyond their immediate vicinity. Generation of nanobubbles addresses the above issues as their smaller sizes allow for increased circulation times, extravasation to less accessible regions, and retention by the enhanced permeability and retention effect (Nomikou and McHale 2012).

Bubble liposomes are echogenic liposomes composed of 1,2-distearoyl-sn-glycero-3-phosphatidyl-ethanolamine-polyethyleneglycol ($DSPE\text{-}PEG_{2000}\text{-}OMe$) and a phospholipid [1,2-dipalmitoyl-sn-glycero-3-phosphocholine (DPPC) or 1,2-distearoyl-sn-glycero-phosphatidylcholine (DSPC)]; they are filled with perfluoropropane gas and have an average size that is less than 1 μm (Fig. 2.9) (Un et al. 2010, 2011a, b; Sugano et al. 2012; Omata et al. 2011, 2012; Suzuki et al. 2010; Endo-Takahashi et al. 2012; Horie et al. 2010, 2011; Kodama et al. 2010). Nucleic acid is mixed with bubble liposomes to generate an acoustically active bubble lipoplex. These bubble lipoplexes can be further functionalized by attaching different ligands [laminin peptide (Negishi et al. 2010), mannose (Un et al. 2010, 2011a, b), and folate (Omata et al. 2012)] for active targeting. Bubble liposomes, and other echogenic liposomes, have several suitable characteristics for enhanced gene delivery: (1) high gene-loading capacities similar to conventional liposomes, (2) active targeting upon functionalization, (3) controllable release, and (4) suitable echogenic properties for sonoporation and image-guided delivery (Huang 2008). Studies by Negishi et al. (2010) and Omata et al. (2011, 2012) have demonstrated co-administration of a PEGylated transfection vector and bubble liposomes with ultrasound to significantly enhance transfection efficiencies when compared to the vector alone; induced oscillations and cavitation by bubble liposomes are postulated to permit endosomal escape, a process generally inhibited as a result of PEGylation (Fig. 2.10). Improved endosomal escape is attributed to transient influx of Ca^{2+} associated with pore formation. As endosomal acidification is influenced by Ca^{2+}, influx of Ca^{2+} affects endosome functionality and contributes to endosomal escape (Omata et al. 2011). It is also feasible to suggest that ultrasound-induced sonophores may also play a role in endosomal escape. Since bubble lipoplexes are similar to conventional lipoplexes, they are capable of transfecting cells in the absence of ultrasound. Studies by Endo-Takahashi et al. (2012) and Un et al. (2011a) have shown that using bubble lipoplexes as both the acoustically active agent and

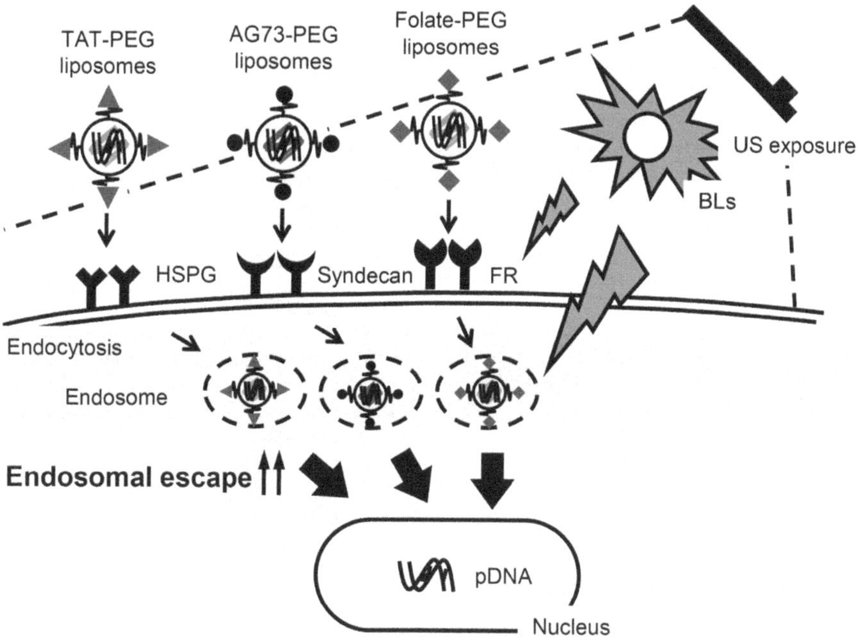

Fig. 2.10 Promotion of intracellular trafficking of liposomes co-administered with bubble liposomes and ultrasound exposure. Adapted from Omata et al. (2011)

transfection vector results in direct gene delivery through pores rather than through endocytosis and endosomal escape. Enhanced transfection efficiencies with bubble liposomes are observed in in vivo gene delivery into tongue tissue (Sugano et al. 2012), bladder (Horie et al. 2010), and skeletal muscle (Kodama et al. 2010) as well as different formulations of cancer gene therapeutics (Un et al. 2010, 2011b; Suzuki et al. 2010; Horie et al. 2011).

Poly(D,L-lactic-co-glycolide) (PLGA) polymers have been considered excellent materials for nanobubble synthesis as they are approved by the FDA. They possess the following properties which are amenable for drug delivery applications: biodegradability, drug biocompatibility, mechanical properties, and ease of processing (Figueiredo and Esenaliev 2012). In addition, solid polymeric nanoparticles are capable of lowering the cavitation threshold, in the absence of gas bubbles, upon ultrasound exposure. Various aspects of PLGA in ultrasound-mediated gene delivery are described in a detailed review by Figueiredo and Esenaliev (2012).

Overall, physical forces such as ultrasound pulsing have been shown to greatly improve cell uptake and tissue penetration of oligonucleotides as delivered with lipid or polymeric carriers. Although the deep tissue penetration of sonoporation or microbubbles remains limited, their more transient and noninvasive nature makes this a highly valuable gene delivery tool, particularly for the delivery of genetic material past difficult barriers such as the BBB.

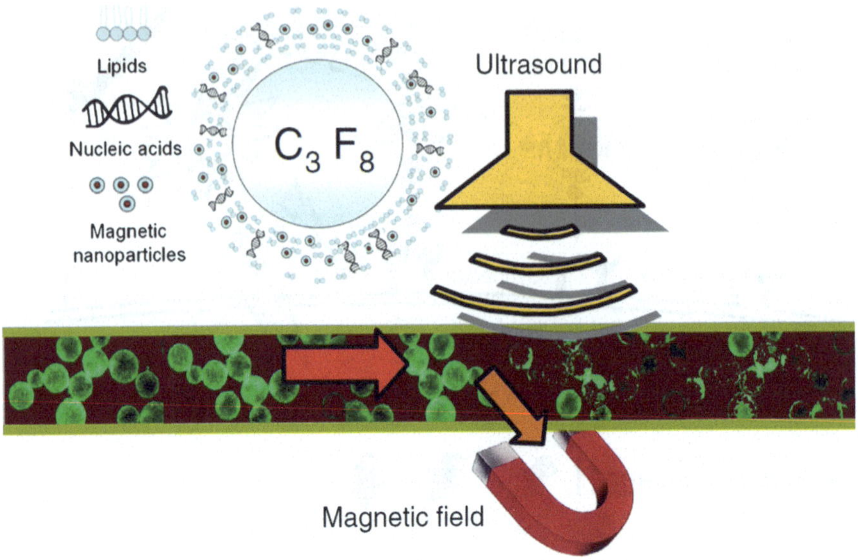

Fig. 2.11 Site-specific targeted delivery mediated by the effects of ultrasound and magnetic field on magnetic microbubbles. Adapted from Plank et al. (2011)

2.6 Combination of Acoustic and Magnetically Active Nanoparticles

Magnetofection and sonoporation both have inherent properties that enhance transfection efficiency (see Sects. 2.4 "Magnetofection" and 2.5 "Sonoporation—Nano Microbubbles and Bubble Liposomes"). Briefly, magnetofection enables rapid sedimentation of magnetic vectors to cell surface; this improves transfection kinetics and promotes close interaction between vector and cell membrane. Sonoporation facilitates internalization through transient pore formation and upregulation of endocytosis. In in vivo settings, magnetic vectors can achieve efficient localization at target sites through the administration of an external magnetic field, while microbubbles promote site-specific delivery upon the administration of ultrasound waves. In addition, both techniques are (1) minimally invasive, (2) are suitable for repeat administrations, and (3) act as stand-alone image modalities enabling diagnosis and image-guided delivery. The effects of magnetofection and sonoporation act in perfect marriage, eliciting synergistic enhancements to transfection efficiency (Fig. 2.11).

Magnetic microbubbles are gene delivery vectors that can be remotely controlled by magnetic fields and ultrasound. The combined effects of magnetofection and sonoporation can be elicited through the co-administration of magnetic nanoparticles (MNP) and nonmagnetic microbubbles along with activation by the application of magnetic field and ultrasound. Alternatively, magnetic microbubbles can be generated by loading MNPs into microbubbles or by attaching MNPs onto the surface of the microbubble (Owen et al. 2012). Through co-administration of L-α-phosphatidylcholine phospholipid microbubbles (~6 μm) and magnetic micelles (~2 μm), Stride et al. (2009) were able to demonstrate increased transfection of naked plasmid DNA in vitro; this assumption is due to sensitization of cells from magnetic micelles and magnetic field exposure. Similarly, the incorporation of magnetic micelles into microbubbles enhances transfection as a result of magnetically induced close proximities between cells and microbubbles (Stride et al. 2009). Mannell et al. (2012) were able to demonstrate efficient binding of plasmid DNA to magnetic nanoparticle-coated DPPC/DPPE phospholipid microbubbles and subsequent enhanced transfection efficiencies in vitro and in vivo; site-specific localization and targeted delivery are confirmed to be mediated by the synergistic effects of magnetofection and sonoporation. The incorporation of all three components (microbubble, MNP, and nucleic acid) has positive implications in in vivo applications as it resolves the issue relating to the differing biodistributions of the three components during co-administration.

Vlaskou et al. (2010a) have demonstrated the generation of magnetic and acoustically active lipospheres (MAALs) comprised of soybean oil, stabilized by Metafectene/1,2-dioleoyl-sn-glycero-3-phosphoethanolamine (DOPE), with different surfactant-coated MNPs and perfluoropropane gas (Tween60-MAG-AAL and PEI-MAG2-AAL). The sequential addition of MNP and nucleic acid results in greater microbubble aggregation than simultaneous mixing of MNP and nucleic acid (Vlaskou et al. 2010a). Simultaneous mixing of MNP and nucleic acid prevented aggregations such that the microbubbles are suitable for systemic delivery; however, this comes at a cost as MAAL-nucleic acid interactions are substantially hampered. Localization of MAALs into specific targeted regions has been confirmed upon application of ultrasound and magnetic field under in vivo conditions; high nucleic acid-binding capacity and magnetically induced close proximities between cells and microbubbles are observed in vitro as evidenced by enhanced transfection efficiency with magnetic field (Vlaskou et al. 2010a). However, MAALs do not possess sufficient acoustic activities to enhance transfection by sonoporation. This paves the way for the development of a second generation of magnetic microbubbles comprising of a DPPC/DSPE-PEG_{2000}-biotin lipid bilayer similar to that of a bubble liposome (Vlaskou et al. 2010b). The incorporation of MNPs and nucleic acids, along with subsequent loading of perfluoropropane gas, results in magnetic microbubbles with sizes of 1–2 μm. Using the second-generation magnetic microbubbles, gene delivery occurs in a plasmid concentration-dependent manner, a feature not observed in nonmagnetic microbubbles (Vlaskou et al. 2010b). In addition, such enhanced gene delivery can only be observed upon ultrasound exposure; hence, these microbubbles are acoustically active and can induce sonoporation. The protocol for the synthesis of MAALs and its second-generation microbubbles is described in detail by Vlaskou et al. (2013).

The physical (size, size distribution, and concentration) and acoustic properties of magnetic microbubbles only vary slightly when compared to nonmagnetic microbubbles of the same formulation (Mulvana et al. 2012). The magnetic properties of each individual microbubble vary depending on the amount of MNPs bound to the microbubble. Application of a magnetic field induces an attractive force that can overcome the shear forces of blood flow and localizes magnetic microbubbles in a confined area (Mannell et al. 2012). Subsequent microbubble disruption by ultrasound increases endothelial permeability, increases extravasation, and enhances gene delivery. The concerted effects allow site-specific targeting with high precision; this reduces the likelihood of adverse effects. Thus, the use of magnetic microbubbles for gene delivery poses to be a novel approach that will be thoroughly explored in the coming years.

2.7 Optical Transfection

Optoinjection, the light-mediated transfer of membrane-impermeable substances, and optical transfection, the process of light-mediated nucleic acid delivery, stem from the induction of membrane permeations and the generation of small transient membrane pores (optoporation) through the administration of continuous wave (CW) or pulsed laser irradiation. There are several inherent advantages in using optical transfection for nucleic acid delivery. The noncontact nature of optical transfection permits an aseptic means of delivery in sterile environments (Praveen et al. 2011; Soman et al. 2011). High precision of laser irradiation enables selective delivery with high specificity and minimal collateral damage. Lastly, optical transfection can further be integrated with microscopy imaging, optical tweezers, and microfluidic systems (Soman et al. 2011; Antkowiak et al. 2010). The combination of optical transfection and microscopy imaging enables simultaneous visualization of nucleic acid delivery and subsequent intracellular trafficking processes. High selectivity and high throughput can be achieved through integration with optical tweezers and microfluidic systems (Kim et al. 2011; Rendall et al. 2012). However, such advantages are overshadowed by insufficient penetration depth, approximately only 2 mm using infrared wavelengths within the optical window, which severely limits in vivo applications of optical transfection (Stevenson et al. 2010).

The proposed mechanism of membrane permeation varies with respect to laser source, wavelength, and pulse duration (Antkowiak et al. 2010; Kim et al. 2011; Stevenson et al. 2010; Yao et al. 2008). The earliest works have employed ultraviolet (UV) light to induce optoporation on the basis of light absorption by membrane components; however, the application of UV light raised concerns relating to irreversible membrane damage, target cell death, as well as potential damage to the administered exogenous DNA (Yao et al. 2008). The implementation of blue CW lasers (488 nm) has been shown to induce transient membrane permeability for the delivery of plasmid DNA through localized heating from light absorption by phenol red, a cell medium component (Kim et al. 2011; Stevenson et al. 2010; Yao et al. 2008). Using this technique, high postirradiation viability and successful stable/transient transfections have been reported, albeit at low transfection

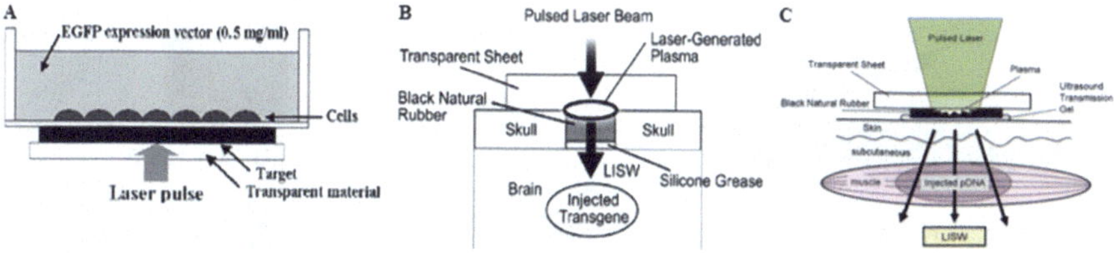

Fig. 2.12 LISW experimental configurations: (**a**) in vitro cell transfections; (**b**) in vivo transfection of mouse CNS; (**c**) in vivo transfection of rat facial muscle. Adapted from Terakawa et al. (2004), Satoh et al. (2005), Kurita et al. (2011)

efficiencies (Stevenson et al. 2010). Although the utilization of blue light reduces significant detrimental effects, potential cellular alterations and deleterious effects may still occur from absorption by various intracellular constituents.

Alternatively, there has been the development of pulsed lasers, whereby the mechanism of membrane permeation and their resultant effects are dependent on pulse duration. The application of nanosecond-pulsed laser generates photomechanical stress waves responsible for transfection by laser-induced stress waves (LISW) (Fig. 2.12). The generation of stress waves induce shear forces that cause transient increase in membrane permeability without affecting viability so long as the applied stress waves are maintained below the damage threshold; this allows entry of exogenous DNA by passive diffusion (Yao et al. 2008). As such, cell viability and transfection efficiencies greatly depend on the distance between cells and the propagation of the laser-induced shockwave. LISW in gene therapy has been effectively applied, in vitro and in vivo, through the implementation of experimental configurations highlighted in Fig. 2.12. Irradiation with 552-nm nanosecond laser pulses from a *Q*-switched Nd:YAG laser, targeting an elastic natural rubber disk (laser target), creates stress waves resulting from the optical breakdown, ablation, or rapid heating of the laser target; stress wave propagation through cells and/or skin on the opposite side induces transient increase in membrane permeability suitable for gene delivery (Yao et al. 2008; Terakawa et al. 2004).

Using this configuration, with careful consideration to several critical parameters (e.g., impulse and number of LISWs), transfection of plasmids encoding luciferase, enhanced green fluorescent protein (EGFP), and/or β-galactosidase have been demonstrated in NIH3T3 murine fibroblast (Terakawa et al. 2004) in vitro as well as rat skin/muscle (Kurita et al. 2011; Ogura et al. 2004) and mouse central nervous system (CNS) (Satoh et al. 2005) in vivo (Fig. 2.13). Irradiation following intradermal injection of plasmid DNA has shown to induce selective expression in rat epidermal cells in vivo (Ogura et al. 2004). In addition, a positive correlation between transgene expression and number of administered pulses (up to three pulses) has been observed; administration of three pulses constitutes an increase in transgene expression corresponding to a 120-fold increase in luciferase activity along with sustained luciferase expression for up to 5 days post-transfection (Ogura et al. 2004). The application of LISW in combination with delivery vectors composed of PEI/DNA polyplexes has yielded efficient and widespread EGFP expression in mouse CNS,

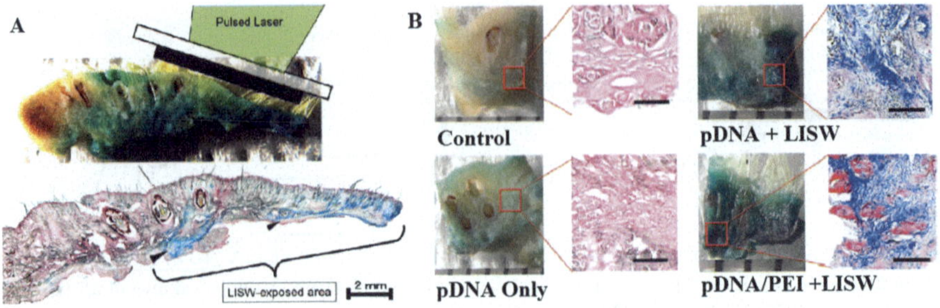

Fig. 2.13 Transgene expression of *lacZ* and β-galactosidase activity after gene delivery into rat facial muscle by LISW: (**a**) Whole facial tissue 7 days post-transfection (scale bar: 2 mm); (**b**) visual comparison between different experimental conditions obtained 7 days post-transfection (scale bar: 200 μm). Adapted from Kurita et al. (2011)

achieving penetration depths of 1.5 mm and 3.5 mm into the cortex surface of adult and newborn mice, respectively (Satoh et al. 2005). Similarly, the combination of LISW and PEI/DNA vectors have resulted in significant increases in transgene (*lacZ* for expression of β-galactosidase) expression as demonstrated by greater β-galactosidase activity in rat facial muscles; β-galactosidase activity has been shown to sustain for up to 14 days with peak activity 7 days post-transfection (Kurita et al. 2011). Gene delivery by LISW therefore enables simultaneous treatment of a large number of cells while achieving targeted delivery through controlled administration of stress waves. In addition, LISW permits deep tissue treatments as stress wave propagation exceeds the limited penetration depth attained from direct irradiation (Terakawa et al. 2004). Lastly, LISW by *Q*-switched Nd:YAG laser offers additional advantages as they are commercially available, cost-effective, easy to operate, safe, and approved for use in head and neck surgeries (Kurita et al. 2011).

In contrast to nanosecond-pulsed lasers, the administration of femtosecond-pulsed lasers, involving near-infrared (NIR) (700–1100 nm) multiphoton processes, has been generally applied for single cell transfections yielding high transfection efficiency and low cytotoxicity (Praveen et al. 2011; Antkowiak et al. 2010; Stevenson et al. 2010; Yao et al. 2008). Direct single target cell irradiation leads to photon absorption and free electron excitation, which contributes to plasma formation and subsequent localized changes in membrane permeability including the formation of cavitation bubbles and submicron transient pores (Praveen et al. 2011; Stevenson et al. 2010; He et al. 2011; Baumgart et al. 2012). Transient pore formation permits the delivery of exogenous DNA in the absence of substantial thermal effects due to ultrashort pulse duration and high repetition rates. Ultrashort pulses prevent heat accumulation, while high repetition rates, greater than the critical repetition rate of 1 MHz, enable the diffusion of excess thermal energy (He et al. 2011). Thermal effects can also be mitigated by using objectives with high numerical aperture (NA). Higher NA objectives are commonly used to focus the laser beam in order to reach high photon density; the resulting small focal volumes permit adequate diffusion of thermal energy into the surroundings. Objectives with NA greater than 1 will exhibit low heat accumulation, whereas objectives with NA of 0.6 or less will exhibit substantial heat accumulation and temperature increase (He et al. 2011).

Femtosecond-pulsed laser-mediated transfections are at a disadvantage due to low throughput, limited loading concentrations, and the requirement for precise instrumentation to achieve direct laser alignment (Yao et al. 2008). Such issues may be mitigated with the implementation of Bessel beams and spatial light modulator (SLM). Laser alignment is crucial for successful transfection as direct membrane interaction can only be achieved upon precise positioning of the beam on the membrane. The use of a Gaussian beam demands precise laser alignment due to a very short effective focal length contributing to a confined axial therapeutic range; such properties raise difficulties in focusing the laser beam on the membrane as any minute misalignments, as little as 3 μm, may result in over 50% reduction in transfection efficiency (Stevenson et al. 2010; He et al. 2011; Tsampoula et al. 2007). Bessel beams, characterized as non-diffracting rods of light, possess an axial therapeutic range that is 20 times greater than that of Gaussian beams, thus overcoming issues relating to misalignment and the requirement for precise beam focus (He et al. 2011; Tsampoula et al. 2007). Since only the central core has sufficient energy to induce free electron excitation and plasma formation, Bessel beams introduce less damaging effects in comparison to out of focus CW light (Stevenson et al. 2010). In combination with SLM for steering and emission of multiple arrays of Bessel beams, the potential for "point and shoot" capabilities may enable simultaneous targeted transfection of preselected groups of cells (Stevenson et al. 2010; Cizmár et al. 2008). As demonstrated by Antkowiak et al. (2010), SLM enables submicron three-dimensional positioning that results in consistent, repeatable poration and time-sequenced irradiation at multiple axial and lateral positions (Fig. 2.14). The application of "point and shoot" targeted transfection has permitted selective co-transfections which resulted in co-expression green fluorescent protein (GFP) and Mito-DsRed (Fig. 2.14). Using unassisted raster scan irradiation, the authors are able to achieve high-throughput transfections at an average rate of one cell per second (Antkowiak et al. 2010).

Previous studies have shown successful in vitro optical transfection of various cell lines with varying transfection efficiencies: Chinese hamster ovary (CHO, 63%), human embryonic kidney (HEK, 52%), mouse neuroblastoma-rat glioma hybrid (NG108-15, 40%), and human neuroblastoma (SK-N-SH, 45%) cells (Dholakia 2010). In contrast, Tirlapur and König reported transfection efficiencies of 100% in CHO and rat-kangaroo epithelial cells (PtK2) in 2002 (Tirlapur and König 2002). The differences in transfection efficiency may lie in the laser configuration and choice of plasmid construct. Of significance, optical transfection has yielded successful delivery into zebrafish embryos and mouse/human embryonic stem cells. Through optical transfection, the delivery of Simian-CMV-EGFP plasmids into blastomeres of early to mid-cleavage stage zebrafish embryos results in wide distribution of EGFP expression along yolk extension, notochord, floor plate, and somites (Kohli et al. 2007). Application of a sub-20 femtosecond infrared laser pulse constitutes 70–80% transfection efficiency for human salivary gland and human pancreatic stem cells with 100% viability and cell division for some of the

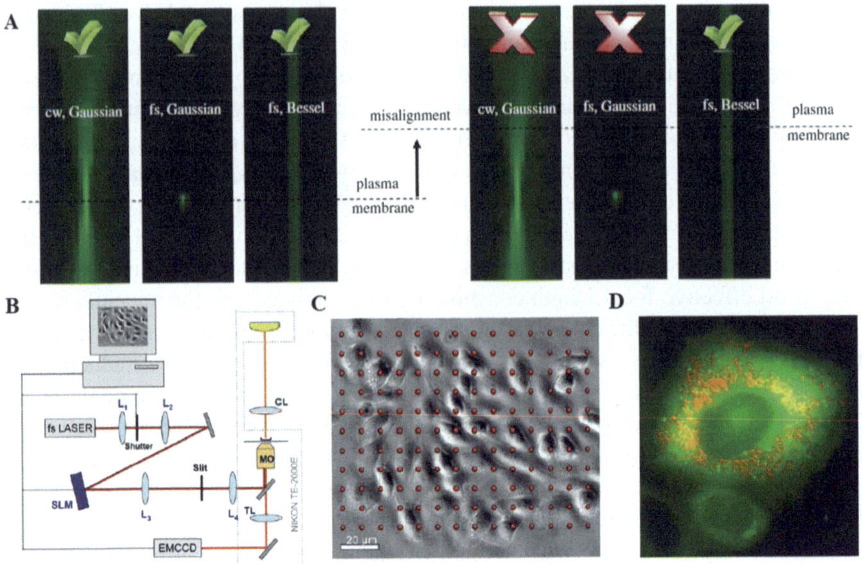

Fig. 2.14 (**a**) Conceptual schematic comparing Gaussian and Bessel beams upon laser misalignment; (**b**) experimental setup using SLM; (**c**) illustration of raster scan irradiation for high-throughput optical transfections; (**d**) co-expression of GFP and Mito-DsRed in Chinese hamster ovary (CHO-K1) cells. Adapted from Antkowiak et al. (2010), Stevenson et al. (2010)

transfected cells (Uchugonova et al. 2008). Optical transfection of mouse embryonic stem cells (E14g2a) with plasmid expressing Mito-DsRed (pDsRed2-Mito) has shown to result in 25% transfection efficiency with demonstrated differentiation potential upon the delivery of endoderm-associated transcription factor Gata-6 gene; development of spindle- and stellate-shaped morphology along with expected up-/downregulation of transcriptional factors, characteristic of extraembryonic endoderm tissue, confirms Gata-6-induced stem cell differentiation (Dholakia 2010). Ishii et al. (2017) recently tracked the intracellular localization of fluorescently labeled plasmid DNA-liposomal complexes (Lipofectamine 3000) as delivered through femtosecond laser pulsation. While DNA uptake occurred regardless of laser irradiation, gene expression was only observed in the event of laser irradiation. They hypothesized that laser irradiation may enhance persistence of the DNA/liposome complex in the endosomal compartment, reducing its degradation in the lysosomal pathway. They further observed increased transport of DNA into the nucleus upon laser irradiation, which they attributed to perforation of the nuclear membrane and subsequent diffusion of DNA/liposome complexes toward the nuclear region. These highly promising results further elucidate the mechanisms by which laser-aided transfections can improve non-viral gene delivery.

Femtosecond-pulsed laser transfections have been applied in vivo for intradermal gene delivery, implicating the potential for long-term gene expression and genetic immunization. As demonstrated by Zeira et al. (2007), intradermal injection of plasmid into mice, conferring luciferase reporter expression, constitutes a 625-fold increase in luciferase expression 43 days after 2-minute irradiation under optimized parameters (780 nm, 200 fs pulses, 76 MHz); luciferase expression remained stable for over 7 months in contrast to a lack of transgene expression for nonirradiated mice. Vaccination by intradermal injection of hepatitis B virus surface antigen (HBsAg) expressing plasmids and femtosecond-pulsed irradiation induce humoral and cellular immune activation, denoted by elevated levels of IgG1, IgG2a, IFN-γ, and IL-4, along with increases in HBsAg-specific titers that contributed to a 215-fold higher titer on day 93 and 22-fold higher titer on day 210. Vaccinated mice have been shown to be immunized against HBsAg-transfected tumor cells with significant reduction in tumor growth over mice without irradiation (Zeira et al. 2007). Immunity is confirmed to be tumor specific as tumor growth retardation is not observed after implantation of untransfected tumor cells.

In conjunction with optical transfection through direct laser-membrane interactions, optical transfections have been achieved with light-absorbing particles including gold and carbon nanoparticles. As demonstrated by Felgner et al. (1987), the application of 100 nm off-resonance spherical gold nanoparticles and low fluence, femtosecond-pulsed irradiation enables plasma formation, nanocavitation, and membrane permeabilization while preventing nanoparticle fragmentation through limited photon absorption. Low fluence laser irradiation ensures nanoparticle-dependent optical transfection, achieving localized gene delivery into WM278 human melanoma cells with 23% transfection efficiency and low gold nanoparticle cytotoxicity (<1%). Similarly, Wu and Wu (1987) achieved a 17-fold increase in luciferase reporter gene expression after gene delivery into DU145 human prostate cancer cells upon femtosecond irradiation of carbon black nanoparticles. However, the mechanism of membrane permeabilization is attributed to a photoacoustic effect producing cavitational shockwaves through light-mediated carbon steam reaction.

In addition to gene delivery by optical transfection, light has been implemented for endosomal escape and activation of RNA interference (RNAi) upon successful internalization of small interfering RNA (siRNA) or small hairpin RNA (shRNA) (Fig. 2.15). Photochemical internalization (PCI) employs light activation of internalized photosensitizers (PS) for reactive oxygen species (ROS) generation that facilitates photochemically induced endosomal escape (Yao et al. 2008; Matsushita-Ishiodori and Ohtsuki 2012). Endocytosis of siRNA complexes with PS permitted cytosolic release for siRNA/shRNA; however, RNAi by PCI must be finely tuned in order to circumvent issues relating to PS toxicity and potential photodynamic effect on other photodynamic therapy (PDT) targets including nuclei, mitochondria, lysosomes, and plasma membrane. CPP-linked RBP-mediated RNA internalization and photoinduced RNAi (CLIP-RNAi) served as an improvement to PCI as the PS is directly linked to shRNA through a photosensitizing carrier consisting of a cell-penetrating peptide, an RNA-binding protein, and a photosensitive fluorescent dye; application of CLIP-RNAi has shown a RNAi efficiency of up to approximately 80% while preventing any un-induced leakage activity observed in PCI (Matsushita-Ishiodori and Ohtsuki 2012; Matsushita-Ishiodori et al. 2011). Gold nanostructures, with strong surface plasmon resonance absorption at NIR

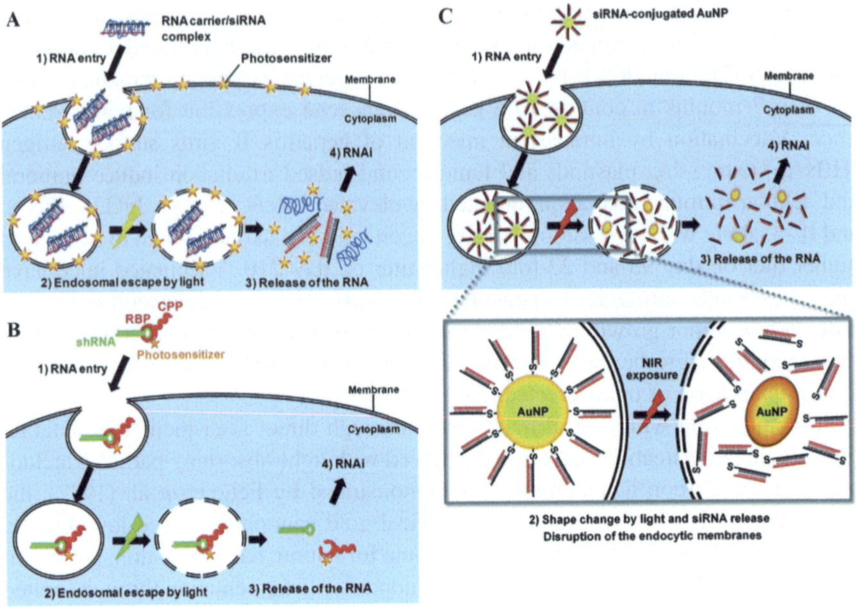

Fig. 2.15 Light-activated RNAi: (**a**) PCI; (**b**) CPP-linked RBP-mediated RNA internalization and CLIP-RNAi; (**c**) photochemical transfection. Adapted from Matsushita-Ishiodori and Ohtsuki (2012)

wavelengths and acceptable in vitro/in vivo toxicities, enable temporal- and spatial-controlled RNAi by photothermal transfection (Matsushita-Ishiodori and Ohtsuki 2012; Braun et al. 2009; Lu et al. 2010a).

As demonstrated by Braun et al. (2009), application of a TAT-lipid-coated gold nanoshell-siRNA conjugate permitted efficient internalization by endocytosis and subsequent endosomal escape by heating, cavitation, and endosome leakage upon femtosecond-pulsed irradiation of gold nanoshells. At the same time, irradiation causes Au plasmon-induced destabilization and cleavage of polyethylene-glycol-thiol bonds for the dissociation of surface-bound anti-EGFP siRNA from gold nanoshell structures, thus achieving cytosolic release and RNAi. Photoinduced dissociation of the siRNA was denoted by controllable increases in fluorescence of siRNA-bound Cy3, a fluorescent molecule that was actively quenched by gold. Through photothermal transfection, the authors are able to induce ~80% reduction in GFP; a lack of cytosolic release of nanoshells corroborates with the lack of GFP downregulation in absence of irradiation (Braun et al. 2009). Using folate functionalized, anti-NF-KB p65 siRNA-bound, hollow gold nanospheres (F-PEG-HAuNS-siRNA), Lu et al. (2010a) achieved selective uptake by HeLa cells overexpressing folate receptors and up to 95% p65 downregulation 72 h after photothermal transfection. The authors attributed the cytosolic release of siRNA to structural transformations of nanospheres, as confirmed by TEM, accompanying siRNA dissociation. In vivo studies, with mice harboring HeLa cervical cancer xenografts, indicate significant uptake of folate functionalized nanospheres and significant downregulation of NF-KB p65 after NIR irradiation. Enhanced apoptotic response, achieved through the combined treatment of p65 siRNA photothermal transfection and irinotecan, contributes to a significant delay in tumor growth (Lu et al. 2010a). Building on the concept of RNAi by photothermal transfection, an elegant system of photonic gene circuits for bimodal photothermal regulation of NF-KB p65 expression has been created; the development of such an intricate system is described in detail by Lee et al. (2012).

2.8 Gold Nanostructures

Gold nanoparticles (GNPs) are attractive for use in drug and gene delivery due to a number of favorable properties such as inertness and low toxicity, straightforward synthesis with low size dispersion, easy functionalization (usually through thiol linkages), and photophysical properties (Ghosh et al. 2008). Depending on the synthetic method used, gold nanoparticle core sizes have been known to range from 1 to 150 nm (Ghosh et al. 2008). In the following section, the use of GNPs to gene delivery will be discussed, with focus on the properties of GNPs that uniquely facilitate such an application.

2.8.1 Gold Nanoparticles as Oligonucleotide Carriers

When used as non-viral gene delivery vectors, strategies to achieve efficient gene delivery often involve chemically functionalizing the gold nanoparticles either with the therapeutic oligonucleotides themselves or with other existing transfection agents to aid in complex formation and delivery (Barhoumi et al. 2009). When DNA is bound to gold nanostructures, steric hindrance increases the DNA half-life from minutes to hours against digestion from nucleases (Barhoumi et al. 2009). Additionally, electrostatic repulsion between polyvalent cations near the gold surface and dications within the nucleases contributes to oligonucleotide stability (Barhoumi et al. 2009).

Chemical conjugation of cationic molecules such as polyethyleneimine (PEI), quaternary ammonium chains, cationic lipids, or cationic amino acids has been used to improve electrostatic interaction with the nucleic acid cargo, thus enabling efficient gene delivery (Cebrián et al. 2011). For example, thiol-modified 2 kDa PEI has been conjugated to gold nanoparticles and used to deliver green fluorescent protein (GFP)-encoding plasmid DNA in vitro (Sharma et al. 2011). Here, GFP expression was seen in nearly 50% of human corneal fibroblasts within 12 h following transfection and cellular phenotype, and viability appeared unaltered with nitrogen/phosphorous ratios up to 180 (Sharma et al. 2011). Topical application of PEI2-GNPs to rabbit corneas in vivo resulted in PEI2-GNPs uptake 12 h later,

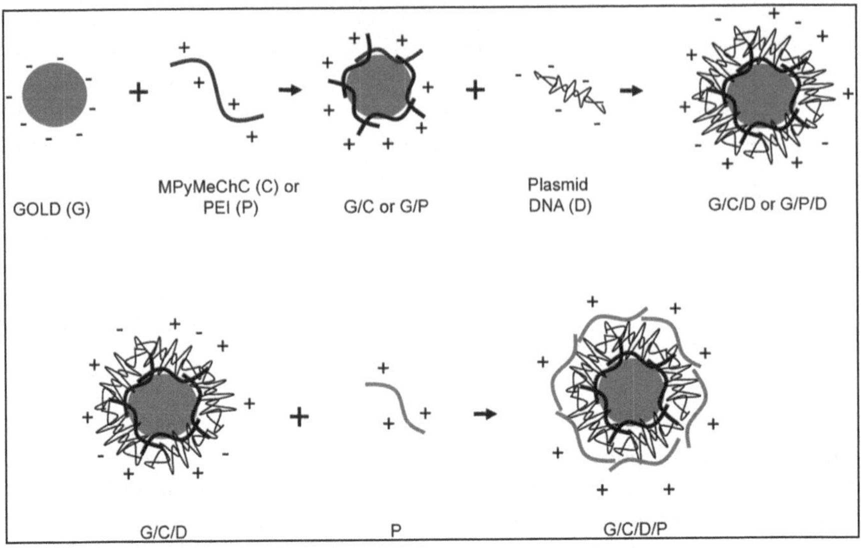

Fig. 2.16 Multilayer gold nanoparticles used in and adapted from Tencomnao et al. (2011)

with an indication of slow clearing from the cornea and no significant immune response compared to untreated corneas (Sharma et al. 2011). Furthermore, transmission electron microscopy data of treated rabbit corneal keratocytes showed PEI2-GNPs within the extracellular matrix and endosomes near the cell surface, as well as evidence of endosomal rupture in corneal keratocytes, suggesting the nanoparticles are taken up by endocytosis and that PEI in PEI-conjugated gold nanoparticles likely continues to facilitate cargo release from the endosome through the "proton sponge" effect seen in other PEI-containing systems (Sharma et al. 2011).

Similarly, gold nanoparticles may be encapsulated by a polymer to improve DNA binding and delivery. GNPs entrapped by amine-terminated generation 5-poly (amidoamine) (PAMAM) dendrimers ($G5.NH_2$) formed smaller complexes with plasmid DNA (pDNA) than $G5.NH_2$ alone and demonstrated lower cytotoxicity and 100 times greater transfection efficiency (Shan et al. 2012). Shan et al. claimed that the entrapped gold nanoparticle helped preserve the 3D spherical morphology of the dendrimer, allowing for more binding sites with DNA (compared to dendrimers without a gold core), higher compaction of the pDNA, smaller particle sizes, and enhanced gene delivery (Shan et al. 2012).

In another scenario, methylated *N*-(4-pyridinylmethyl) chitosan chloride (MPyMeChC) and PEI have also been used to generate multilayer gold-polymer-DNA nanoparticles (Tencomnao et al. 2011). Figure 2.16 shows an illustration of the nanoparticles used by Tencomnao et al. in 2011. Use of the gold nanoparticle scaffold in this study enhanced transfection efficiencies compared to the C/D and P/D polyplex systems (Tencomnao et al. 2011). Furthermore, combining gold and PEI resulted in transfection efficiency 10 times greater than that of DNA with PEI (P/D) alone; G/C/D/P and G/P/D complexes produced results comparable to

Lipofectamine 2000 in HeLa and A549 cell lines (Tencomnao et al. 2011). PEI-containing multilayer nanoparticles (G/C/D/P and G/P/D) showed higher efficiency than that of G/C/D systems, seemingly due to the proton sponge effect brought along by the PEI content (Tencomnao et al. 2011). Additionally, the investigators claimed that the use of colloidal gold improved the monodispersity of the carrier samples as well as facilitated easier release of the DNA after cell internalization compared to polyplex systems where DNA is tightly condensed (Tencomnao et al. 2011). In cell viability assays, toxicity increased with the introduction of gold (60–80% viability) versus C/D and Lipofectamine (80%); however, the toxicity remained lower than that of P/D complexes (~50% viability) (Tencomnao et al. 2011).

2.8.2 Photophysical Properties of Gold Nanoparticles and Gene Delivery

Alternatively, a number of groups have exploited the photophysical properties of gold nanoparticles and nanorods, namely, the surface plasmon resonance, to achieve remotely triggered gene delivery to target cells. In brief, exposure to an electromagnetic frequency prompts a collective oscillation of the valence electrons at the surface of spherical nanoparticles with a diameter much smaller than the light wavelength (Boisselier and Astruc 2009). This oscillation is known as the surface plasmon resonance, which is strongest for noble metals (gold, silver, and copper) and is shifted into the visible portion of the electromagnetic spectrum for these metals (Boisselier and Astruc 2009; Link and El-Sayed 2003). The width of the surface plasmon absorption and frequency depend on the nanoparticle size and shape, as well as the dielectric constant of the metal and the surrounding medium (Link and El-Sayed 2003). Small nanoparticles (<100 nm) absorb light and display this surface plasmon resonance; meanwhile as the particle size increases to the micron scale, the particle becomes a better light scatterer than light absorber (Barhoumi et al. 2009; Boisselier and Astruc 2009). Therefore manipulation of the nanoparticle shape and size allows its resonance to be tuned to a desired resonant frequency (Barhoumi et al. 2009; Boisselier and Astruc 2009). For example, nanoparticles with optical resonances in the near-infrared (NIR) range, where blood and tissue are nearly transparent, are especially useful for biomedical applications (Barhoumi et al. 2009; Boisselier and Astruc 2009). Furthermore, upon illumination, plasmon-resonant nanoparticles dissipate energy through their lattice phonons, converting light to heat (Barhoumi et al. 2009). Thus, the nanoparticle's photothermal response may be used for applications such as photothermal tumor ablation or light-assisted biomolecule delivery like those discussed below (Barhoumi et al. 2009).

Light-induced, specifically NIR-induced, release of DNA or RNA cargo is a promising method of achieving controlled release for gene delivery. In 2009,

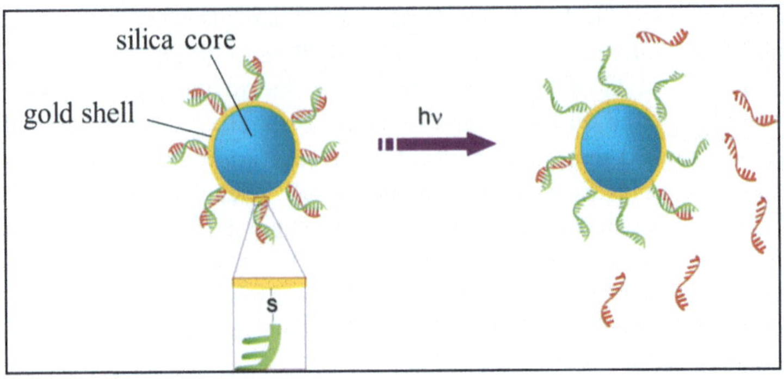

Fig. 2.17 Gold nanoshells for NIR-induced ssDNA release. Adapted from Barhoumi et al. (2009)

nanoshells consisting of a spherical silica core and a gold shell with a plasmon frequency determined by its inner and outer radii were used to demonstrate light-induced release of single-stranded DNA (ssDNA) for antisense therapy (Fig. 2.17) (Barhoumi et al. 2009). In this case, double-stranded DNA (dsDNA) was covalently bound to the gold shell via a 5′-thiol group present only on the sense strand, leaving the complementary strand (the therapeutic antisense oligonucleotide) free to be released from the nanostructure upon thermally induced or NIR light-induced DNA melting (Barhoumi et al. 2009). Light-induced DNA release occurred near-ambient temperature in solution (minimal temperature increase was observed) and appeared to be independent of length for oligonucleotides in the range of 20–70 base pairs (Barhoumi et al. 2009; Huschka et al. 2011). The mechanism for dehybridization under these conditions appears to be due to hot electrons (from plasmon decay) transferred to the adsorbed DNA leading to dehybridization due to repulsion between the two strands (determined in a subsequent study) (Barhoumi et al. 2009; Huschka et al. 2011). On the other hand, thermally induced release was attained at the DNA melting temperature, which is influenced by the nucleotide sequence, type of linkage to the nanoparticle surface, and solution properties such as ion concentration (Barhoumi et al. 2009; Huschka et al. 2011). There was also a notable difference in release efficiency between the two methods; about 50% of the available ssDNA was released with light, but virtually all of the ssDNA was released using the thermal method (Barhoumi et al. 2009; Huschka et al. 2011).

The nanoshell-dsDNA complex was also used for NIR light-triggered release of 4′,6-diamidino-2-phenylindole (DAPI) in H1299 lung cancer cells in vitro (Huschka et al. 2010). Here it was demonstrated that intracellular light-triggered release can be achieved and visualized with a DNA-binding fluorescent molecule such as DAPI and that the complex and the laser power and irradiation time does not adversely affect cell viability (Huschka et al. 2010).

On the other hand, Lukianova-Hleb et al. took a different approach in using the light-to-heat converting property of plasmon-resonant GNPs to achieve cell-specific gene delivery within heterogeneous cell samples. Gold spheres 60 nm in diameter were conjugated with target-specific antibodies and added to the cells, accumulating

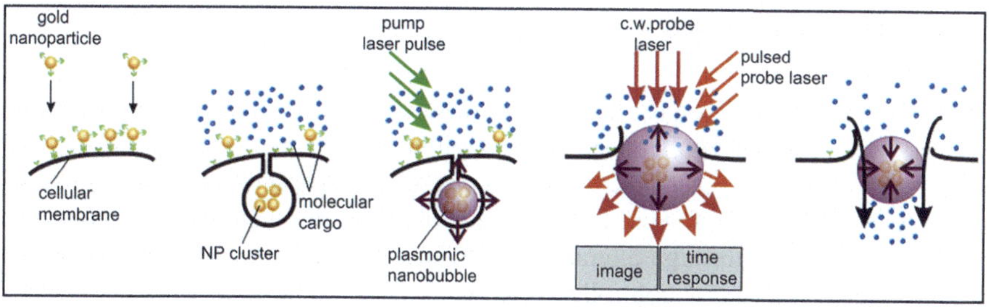

Fig. 2.18 The principle of nano-injection of extracellular cargo via formation of plasmonic nanobubbles. Adapted from Lukianova-Hleb et al. (2012)

in target cells more than nontarget cells through receptor-mediated endocytosis (Lukianova-Hleb et al. 2012). Cell-specific delivery was described as follows (Fig. 2.18). When exposed to a laser pulse, the optical energy was converted to thermal energy through plasmon resonance of the gold nanoparticle clusters. Local heat released by the GNPs evaporated the liquid nanoenvironment, resulting in the generation of transient vapor nanobubbles, known as plasmonic nanobubbles (PNBs). In this case, only minimal optical energy was supplied by the pump laser pulse so that only large clusters of GNPs (mostly in target cells) are capable of forming PNBs at such a low energy; PNB generation was therefore prevented around the smaller clusters and single nanoparticles present in nontarget cells (Lukianova-Hleb et al. 2012). The expanding PNBs formed a transient and reversible nanohole in the cell membrane, and upon PNB collapse, an in-bound jet injected the extracellular matrix including the cargo into the cytoplasm. The investigators reported large nanoparticle clusters in 99% of target cells and successful (96%) target-specific transfection with a GFP-encoding plasmid in the heterogeneous cell sample of CD3-positive human T cells (target) and CD3-negative cells (nontarget) after a single pulse flow treatment along with 75% viability in treated target cells (Lukianova-Hleb et al. 2012).

The reader is referred to the following materials for examples of similar techniques in gold nanoparticle-mediated delivery of other nucleic acid structures such as small interfering RNA (siRNA), microRNA, and short hairpin RNA (shRNA): Lu et al. (2010a), Ghosh et al. (2013), Kong et al. (2012b), Conde et al. (2012), Ryou et al. (2011), Zheng et al. (2012).

A final application of gold nanoparticles is their use as templates in the development of spherical nucleic acid structures which are capable of cellular internalization and gene regulation (Young et al. 2012; Cutler et al. 2012). These structures traditionally consist of gold cores with DNA shells, but investigations have shown that cross-linking the DNA at the base and subsequently dissolving the gold core produces a hollow structure with many of the key properties of the gold nanoparticle conjugates without concerns of toxicity from the gold core (Young et al. 2012; Cutler et al. 2012).

2.8.3 Gold Nanorods

Similarly, gold nanorods have attracted a considerable amount of attention in gene delivery among other applications such as photothermal therapy, biosensing, and imaging, due to their near-infrared light-enhanced absorption and plasmon resonance (Cui et al. 2011). Owing to the surface plasmon resonance of gold nanorods being easily tunable through alterations in their aspect ratio, gold nanorods can be used for selective and/or controlled release of DNA molecules attached to their surface in the same manner described above for gold nanoshells.

For example, by exploiting the association between gold nanorod aspect ratios and their corresponding surface plasmon resonance, one group was able to demonstrate selective release of two differentially labeled "thiolated DNA 40mers" in a mixed sample of DNA-conjugated gold nanorods (Wijaya et al. 2009). Thiolated 6-carboxyfluorescein-labeled oligomers were conjugated to short, "nanocapsule," nanorods that had a longitudinal surface plasmon resonance at 800 nm; meanwhile, thiolated tetramethylrhodamine-labeled oligomers were conjugated to long, bone-shaped nanorods with a longitudinal surface plasmon resonance at 1100 nm (Wijaya et al. 2009). With this method, when a mixture of the purified, conjugated nanorods was exposed to laser irradiation at one of the surface plasmon resonance wavelengths, only the corresponding nanorod was excited and released its DNA cargo from the surface (50–80% DNA release), leaving the other essentially unchanged (less than 10% DNA release) and the DNA still functional (Wijaya et al. 2009). The release was also tunable by laser fluence (Wijaya et al. 2009). This type of system has potential for multidrug/gene delivery.

On the other hand, in a 2011 study comparing thermal and light-mediated release of ssDNA from a gold nanorod ($[w,l] = [13,47]$) surface, no light-induced release was achieved upon excitation of either the transverse or the longitudinal plasmon resonance (Huschka et al. 2011). This was evidenced by the resultant ssDNA release curves, which were very similar to the thermal release curves and showed no significant increases in release at temperatures below the DNA melting temperature (Huschka et al. 2011). Huschka et al. suggested that light-triggered release occurs through a nonthermal mechanism where hot electrons generated by plasmon excitation of the gold nanostructure are transferred to the adsorbed cargo, thus facilitating DNA dehybridization well below its melting temperature (Huschka et al. 2011). Therefore, the capacity for light-triggered release is affected by the absorption cross section and DNA densities on the nanorod surfaces (Huschka et al. 2011). The calculated near-field intensity enhancements were large for nanorod longitudinal excitation, but this local field would be confined to the tips of the rod and decay rapidly with increasing distance from the gold surface (Huschka et al. 2011). Consequently, if charge transfer is higher in regions of large local field, it is possible that an insufficient amount of DNA was susceptible to charge transfer, resulting in a negative result under the experimental conditions (Huschka et al. 2011).

However, it should be noted that the methods discussed in this section have yet to be successfully demonstrated to any large extent in biological systems. The biosafety of gold nanoparticles has already been demonstrated elsewhere, most notably in clinical trials against prostate cancer (Stern et al. 2016). A recent review by Pedrosa et al. (2015) further highlights investigations into GNPs in preclinical and clinical phases. Additionally, a preliminary study of layered polyelectrolyte-conjugated gold nanoparticles demonstrated encouraging results in cultured prostate cancer cells (Huang et al. 2009). Oligonucleotide-conjugated nanorods were used for optically activated release of antisense oligonucleotides for mRNA interference in breast carcinoma cells in vitro as well (Lee et al. 2009). More recently, a series of transfection studies was carried out examining PEGylated PEI-GNP complexes and their ability to transfect HeLa and HEK 293T cell lines with plasmids of different sizes (3000 bp to 40 kbp) (Mar Encabo-Berzosa et al. 2017). In comparison to a standard transfection reagent (Lipofectamine 2000), their PEI-GNP complex performed admirably; transfection efficiencies from gold-mediated gene delivery even surpassed that of Lipofectamine for the larger DNA (40 kbp). Their studies support the combinatorial approach of polymeric or liposomal gene carriers in conjunction with gold nanoparticles as a viable gene delivery vehicle.

2.9 Carbon Nanostructures for Gene Delivery

Carbon nanostructures hold great promise in their potential as nanoscale biomaterials. Some carbon nanostructures have been deemed useful in gene delivery applications, each with their own favorable properties. The following discussion will be organized by structure type, highlighting particular structures found favorable for gene and drug delivery.

2.9.1 Carbon Nanotubes

Carbon nanotubes (CNTs) are well suited for use in the field of nanomedicine given their high aspect ratios, electronic and optical properties, and large surface area, which can be functionalized and loaded with nucleic acid, drug, or protein therapeutics (Cheung et al. 2010; Varkouhi et al. 2011). Moreover, there have been claims that CNT systems enter cells through a nanoneedle, endocytosis-independent mechanism, possibly contributing to the enhanced transfection results seen with some CNT-mediated delivery systems. However, a definite conclusion regarding the cellular uptake mechanism of CNTs has not been made since groups in the field have reported varying results. In one corner lies the work of Dai et al. (Shi Kam et al. 2004; Kam and Dai 2005; Kam et al. 2006) who, through systematic uptake inhibition assays, demonstrated that energy-dependent endocytosis, specifically through clathrin-coated pits, is essential for cellular internalization of short, well-dispersed protein- or DNA-bound single-walled CNTs. Investigations using the near-infrared fluorescence and Raman properties of single-walled CNTs by Jin et al. (2009) also support this view. In contrast, results from Kostarelos et al. (2007), Lacerda et al. (2012) and Pantarotto et al. (2004) lie in the opposing corner, where CNTs are thought to enter cells independent of energy-dependent endocytosis. A series of CNTs functionalized with various small molecules appeared to enter cells regardless of the functional group, cell type and ability for endocytosis, and presence of endocytosis inhibitors. Given these observations, the cylindrical shape of CNTs and their high aspect ratio, some groups (Cai et al. 2005; Rojas-Chapana et al. 2005; Kostarelos et al. 2007) proposed that CNTs penetrate the cell membrane, like a "nanosyringe." Such notable discrepancies between groups may be due to significant differences in the CNT characteristics: functionalization with macromolecules such as proteins and DNA versus smaller molecules, as well as CNT length (Cheung et al. 2010; Kang et al. 2010). Perhaps different cellular entry mechanisms will dominate depending on the physical properties of the CNTs in question.

For a summary of CNTs for gene delivery to date, the reader is referred to some excellent reviews elsewhere (Cheung et al. 2010; Marchesan et al. 2015). Similar to gold nanostructure delivery systems, investigations have involved surface-modified CNTs for delivery of nucleic acid cargo to cells in vitro. When left "as-is" CNTs are water-insoluble; therefore, chemical modification is necessary for their use in gene delivery applications (Liu et al. 2011b). This not only allows for greater solubility and individualization in hydrophilic environments, but the cationic surface modifications also provide a necessary positive surface charge for interaction with cellular membranes and nucleic acid binding. Binding can occur either through electrostatic interactions or chemical conjugation, depending on the functionalization used.

A number of functionalized CNTs have been designed for nucleic acid delivery, including those functionalized with dendrimer polymers. (Poly(Lys:Phe))-coated double-walled CNTs (DWCNTs) were conjugated to thiol-modified siRNA targeted to survivin as a demonstration of an anticancer non-viral siRNA strategy (Neves et al. 2012). CNT-delivered siRNA demonstrated successful transfection in vitro and exhibited similar levels of survivin knockdown and apoptosis as compared with DharmaFECT delivery of siRNA. Furthermore, the DWCNTs do not show any accumulation within cells after transfection and do not induce cellular stress as measured by mitogen-activated protein kinase phosphorylation. Work by Liu et al. using multi-walled carbon nanotubes (MWCNTs) conjugated to linear polyamidoamine demonstrated lower cytotoxicity and produced in vitro transfection results comparable or higher than that of the polyamidoamine and branched PEI (25 kDa) controls, with little serum sensitivity (Liu et al. 2011b). Intracellular trafficking results with Cy3-labeled pGL3 indicated greater nuclear localization after 24 h. A study reported by Podesta et al. employed amino-functionalized MWCNTs to deliver a toxic siRNA sequence to cells in vitro and in vivo through intratumoral injection into xenograft tumors (Podesta et al. 2009). In vitro transfection with the functionalized MWCNTs (compared to 72 h with the liposomal control) and significant tumor reduction was observed in vivo, corresponding to prolonged animal survival. More recently, MWCNTs were modified with amine groups to deliver siRNA against PLK1 (anti-lung cancer target) in tumor xenografts against a liposomal control, demonstrating significant siRNA retention and uptake in tumors after local administration (Guo et al. 2015).

Alternatively, carbon nanotubes might also be magnetized for improved delivery. Cai et al. demonstrated delivery of and expression of plasmid DNA via magnetically directed CNT penetration ("nanotube spearing") into target cells (Cai et al. 2005). Similar to physical methods of gene delivery, the plasmid DNA bypasses the endocytic pathway and may even penetrate pass the nuclear membrane, potentially increasing transfection efficiency in slow- or nondividing cells. CNTs carrying ferromagnetic nickel particles were sensitized to magnetic agitation. Within a magnetic field, the CNTs complexed with pEGFP-c1 could then be oriented and directed to spear cell membranes in both dividing (Bal17) and nondividing (primary neuron) mammalian cells. Transfection efficiency, as observed by EGFP expression, was observed to be almost 100% after 24 h in rapidly dividing cells and approximately 80% after 48 h in nondividing cells (Cai et al. 2005). More recent work by Jie Chen and colleagues also demonstrated efficient uptake of fluorescein isothiocyanate (FITC)-labeled magnetic carbon nanotubes (mCNTs) into human monocytic leukemia THP-1 cells in vitro (Gul-Uludag et al. 2012). After addition to the cells in culture, samples were exposed to external rotating then static magnetic fields for 10 min. Compared to spherical magnetic nanoparticles, the use of mCNTs resulted in more rapid and efficient uptake. One hundred percent uptake was achieved after 1 h (determined through fluorescence-activated cell sorting) with no decrease in uptake efficiency after 24 and 48 h. Additionally, three-dimensional reconstruction images of confocal microscopy results showed nuclear localization 6 h after delivery with no change in cell viability. The cylindrical shape and magnetically functionalized tip of the mCNTs is thought to allow the mCNT to follow the guidance of an external magnetic field better than spherical magnetic nanoparticles. More favorable properties such as a smaller diameter (~1 nm versus 5–500 nm for nanoparticles) and elongated shape (thought to reduce drag forces) are also thought to enhance delivery. In the future, this sort of method might be used for ex vivo transfection and/or magnetic trafficking to some sites of interest in in vivo administration. For example, it may be possible to deliver mCNTs into monocytes followed by in vivo administration of the monocytes and then apply a clinically appropriate magnetic field to direct the monocytes to a tumor site, thereby using the genetically engineered monocytes as cellular delivery vehicles for cancer gene therapy.

2.9.2 Fullerene

The use of C_{60} fullerene (Kroto et al. 1985), also known as buckminsterfullerene or buckyball, has resulted in successful delivery of exogenous nucleic acids to cells. Although other forms of fullerene exist, for example, C_{70}, fullerene-mediated gene delivery has focused on the C_{60} allotrope. Following early investigations of its interactions with DNA pioneered by Nakamura and co-workers, including the ability of fullerene to oxidatively cleave the DNA duplex through photo-excitation

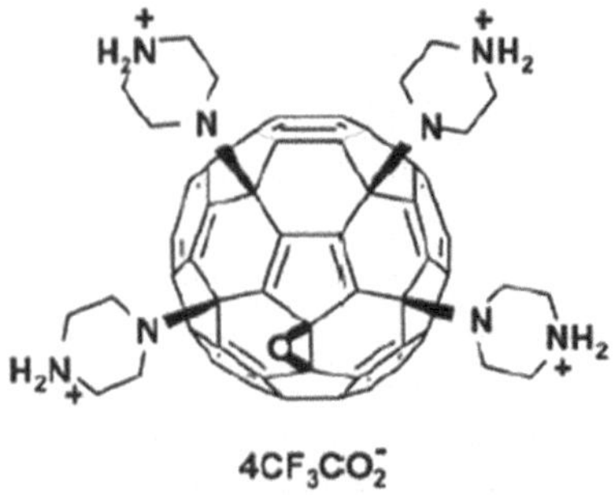

Fig. 2.19 Tetra(piperazino) fullerene epoxide (TPFE), adapted from Maeda-Mamiya et al. (2010)

(Tokuyama et al. 1993), studies have now demonstrated that fullerene derivatives are capable of oligonucleotide delivery into cells (Nakamura et al. 2000; Klumpp et al. 2007; Sigwalt et al. 2011). For more discussion of early studies regarding these initial fullerene derivatives and their structure-activity relationships for gene delivery, the reader is referred to reports and reviews by Nakamura et al. and other groups (Nakamura et al. 2000; Nakamura and Isobe 2003; Isobe et al. 2006a, b; Bakry et al. 2007).

Like the carbon nanotubes, fullerene requires hydrophilic surface modifications for solubility under physiological conditions as an alternative to the use of solvents which often result in cytotoxicity (Nakamura et al. 2000; Isobe et al. 2006b). The resulting amphiphilic fullerene derivatives possess the ability to self-assemble into vesicles, often referred to as "buckysomes" approximating anywhere from 10 to 200 μM in diameter (Partha et al. 2007), and have been used for drug delivery (Partha et al. 2008). Such modifications also allow for DNA complexation and compaction mainly due to electrostatic and hydrophobic forces (Sitharaman et al. 2008). It is notable, as with carbon nanotubes, that a positive charge is necessary for full DNA interaction and complexation.

A cationic functionalized fullerene, tetra(piperazino)fullerene epoxide (TPFE), shown in Fig. 2.19, formed stable complexes with DNA and demonstrated efficient gene delivery in vivo (Maeda-Mamiya et al. 2010). TPFE-DNA complexes remained under 100 nm in serum-containing buffer and achieved more efficient DNA delivery to the liver and spleen compared to Lipofectin without acute liver or kidney toxicity (Maeda-Mamiya et al. 2010). Previously, this system showed successful transfection in vitro and protection from nucleases and has the potential for DNA release through loss or neutralization of the DNA-binding amino groups (Isobe et al. 2006a; Maeda-Mamiya et al. 2010).

Likewise, water-soluble C_{60} derivatives prepared via the Hirsch-Bingel reaction, a simple, high-yielding and versatile fullerene functionalization method, all showed significant in vitro plasmid DNA transfection compared to naked DNA (Sitharaman et al. 2008). However, only two cationic fullerene derivatives, octa-amino and dodeca-amino functionalized fullerenes, at their optimal fullerene-to-DNA-base pair ratio achieved transfection efficiencies comparable to the 25 kDa PEI and CytoPure reagent positive controls. In the presence of serum-free DMEM cell culture medium, the hydrodynamic radii (R_h) of both the octa-amino and dodeca-amino derivatives were below 200 nm, a diameter upper limit that has been associated with more efficient cellular uptake. This may speak to the increased capacity for DNA compaction due to an increased density of positive charges on the surface of these particular derivatives. These samples were shown to consist of polydisperse aggregate populations, which may not be correctly represented in the R_h data (Sitharaman et al. 2008). Additionally, these particular formulations resulted in approximately 30% cell viability 24 h post-transfection and cell morphological changes in culture (no negative control data shown). This cytotoxicity is thought to be largely due to the aggregation of the complexes over time and their subsequent precipitation onto the cell surface. The authors postulate that preventing or disassembling aggregates during complex preparation could also reduce this resultant toxicity. It is also worth noting that although such low cell viability appears highly unfavorable at first glance, these results may be acceptable as dependent on their intended use (e.g., cytotoxicity in a cancer cell population would be quite acceptable and, in fact, desirable) and when keeping in mind that it may not necessarily reflect systemic toxicity (Sitharaman et al. 2008).

Polycationic fullerene hexakis-adducts ($C_{60}(NH_3^+)_{12}$) (Richardson et al. 2000) satisfy the criteria for DNA complexation and cell penetration as these amphiphilic structures possess both positive charges for DNA interaction and hydrophobic regions for cell membrane interaction. Though previously thought unsuitable for transfection due to the isotropic distribution of positive charges (Isobe et al. 2006b), a series of amphiphilic globular fullerene derivatives was demonstrated to have noteworthy gene delivery capabilities, achieving gene expression comparable to a commercial lipid control (JetSI-ENDO Polyplus-Transfection) (Sigwalt et al. 2011). Fullerene hexakis-adduct derivatives were mixed with DNA, and DNA condensation was allowed to occur in 5% glucose, resulting in compact “donut-like” structures less than 100 nm in diameter. The derivatives were similarly mixed in 150 mM NaCl, resulting in aggregated spherical complexes that were observed to mimic PEI-DNA complexes. Transfection efficiency was measured through luciferase expression from delivery of plasmid pCMV-Luc into HeLa cells. In addition to comparable transfection efficiency to the commercial control, cell toxicity was also low. These highly promising results indicate the potential of fullerene hexakis-adducts as nontoxic and efficient transfection agents (Sigwalt et al. 2011).

2.9.3 *Carbon Nanohorns*

A variant of single-wall carbon nanotubes, this carbon nanomaterial was first observed after CO_2 laser ablation of graphene sheets and characterized by Iijima et al. (1999) and Bandow et al. (2000). The single-wall carbon “nanohorn” was thusly named due to the resultant cone-shaped morphology of the resultant particles, which were also observed to assemble in larger “dahlia flower”-like aggregates (requiring purification into individual nanohorns thereafter). The dahlia flower aggregates themselves typically have diameters around 100 nm and are comprised of tightly clustered nanohorns bound by van der Waals forces. Interstitial spaces and the hollow inner "horn" spaces in dahlia aggregates can encapsulate therapeutic cargo as is. To demonstrate their capacity for cargo, dahlia structures were shown to carry and deliver vancomycin hydrochloride in a controlled release manner (Xu et al. 2008).

Nanohorns have great potential as drug carriers due to their ease of synthesis and overall high production yield. As they do not require metal catalysis (Iijima et al. 1999) for synthesis, carbon nanohorns were thought to be less toxic. The extensive surface area inside the "horn" facilitates adsorption of drug cargo and protection from degradation during delivery, allowing it to function much like a nanocage. Murata et al. induced the formation of micropores and mesopores in nanohorn structures with heat (>693 K) (Murata et al. 2001). The micro- or meso-porous structure can further improve the adsorption and release of cargo molecules. Nanohorns and functionalized nanohorns were quickly demonstrated to carry a range of drugs including the aforementioned vancomycin, dexamethasone (Murakami et al. 2004), doxorubicin (Murakami et al. 2006), and cisplatin (Ajima et al. 2008). Fan et al. observed localization of nanohorns within the perinuclear regions of HeLa cells 4–7 h after cell uptake, indicating an endocytic mechanism of cell entry (Fan et al. 2007). They failed to observe nanohorn entry into the nuclear compartment, indicating any nuclear-targeted cargo must dissociate from the nanohorn in order to bypass the nuclear membrane. Miyawaki et al. also investigated the toxicity of carbon nanohorns and concluded low acute toxicities over a 90-day period with rat lung tissue (Miyawaki et al. 2008).

The ability of carbon nanohorns to encapsulate and release drug cargo is very promising. Its intracellular localization without undue toxicity further underlines its prospective use as a gene carrier although investigations into its role as such remain scarce. Guerra et al. reported functionalization of carbon nanohorns with PAMAM dendrimers in order to enhance their solubility and cell penetration (Guerra et al. 2014). PAMAM dendrimers were coupled to the nanohorn via an amidation reaction and were marked with gold nanoparticles to enable tracking and visualization of the hybrid complex. Coupling with the carbon nanohorn enables dispersion of the positive charges across the PAMAM surface, which was thought to reduce its overall toxicity. PAMAM-nanohorn delivery of siRNA targeted against p42-MAPK and GADPH (anti-prostate cancer targets) was evaluated in prostate cancer cell lines (PC-3). The hybrid vectors were indeed able to abolish target gene expression without significant toxicity. The synthesis of these vectors lays the groundwork for future nanohorn-encapsulated oligonucleotide delivery applications, though it remains to be seen if dsDNA can also be delivered with similar success.

2.9.4 Graphene

Graphene is formed of two-dimensional planar sheets of carbon atoms arranged in a honeycomb structure. With its large surface area to facilitate efficient loading of therapeutic cargo and excellent mechanical and chemical properties (Ferrari et al. 2015), explorations into graphene biomedical applications have exploded in recent years. The reader is directed to an extensive review by Ferrari et al. for more information on graphene production and applications (Ferrari et al. 2015).

Graphene does not intrinsically bind nucleic acid as its overall negative charge repulses the negative charge of the phosphate backbone. In 2010, Lu et al. demonstrated the ability of graphene oxide (GO) to bind to and protect oligonucleotides from enzymatic degradation (Lu et al. 2010b). Wu et al. further investigated the extent of DNA adsorption and desorption on GO (Wu et al. 2011). They found that shorter oligonucleotide length, lower pH, higher ionic strength, and lower temperature benefited DNA adsorption, while the reverse was true for desorption. Furthermore, single-stranded DNA bound reversibly to GO and could exchange with free DNA in solution. The strong affinity of GO to short oligonucleotides makes it particularly promising for the delivery of DNA/RNA aptamers or probes (Wang et al. 2010, 2013). Graphene derivatives seem able to bind effectively to single-stranded nucleic acids through π-π interactions but tend to demonstrate weaker dsDNA interactions (Tang et al. 2010a, b; Wen et al. 2010). The binding affinity of functionalized graphene for oligonucleotides can be exploited for biosensor and imaging applications.

In terms of graphene-mediated transfection, Feng et al. constructed hybrid polymers consisting of the gold standard PEI and GO (Feng et al. 2011). Functionalization of GO with PEI enriches the complex in positive charges allowing loading of dsDNA. The GO-PEI complex exhibited greatly reduced cytotoxicity in comparison with bare PEI polymers, even with very high molecular weight PEI (10 kDa), and did not decrease transfection efficiency. Addition of PEG to PEI-GO hybrid vectors further improved transfection efficiency (Feng et al. 2013). PEI-GO conjugates were also further transfected into zebrafish embryos with sustained gene transfection up to 72 h (Zhou et al. 2012). Zhou et al. concluded that reduced free PEI and smaller GO loads ("ultrasmall" GO with diameters approximately 10–20 nm) contributed to the high transfection efficiency. Cationic graphene oxide complexes could very well lead the way to an overall safer and more efficient non-viral gene therapy.

2.9.5 Conclusions

The family of carbon-based nanomaterials far exceeds what has been described in this chapter. However, extensive investigation into toxicity and bioaccumulation effects is necessary before their use in clinical gene therapy studies. Beyond their utility for gene delivery, carbon nanostructures have potential for innumerable applications as the building blocks for nanodevices and other applications. Overall, carbon structures require functionalization with charged constituents for solubilization in aqueous solution. Aggregation with serum components will be an obstacle to delivery in the future, though it may be mitigated with further complexation with

PEG. Thus far, carbon nanotubes, fullerenes, and graphene functionalized with cationic polymers or lipids show great promise for oligonucleotide delivery.

2.10 Localized Hyperthermia

Localized hyperthermia has long served as an effective cancer therapeutic by acting synergistically with chemotherapy and radiotherapy; it has extended the long-term survival of patients suffering from soft tissue sarcoma, melanoma, and recurrent breast/cervical cancers among others (Sanches et al. 2011; Mallory et al. 2016). Localized hyperthermia induces a number of desirable therapeutic effects: direct thermal toxicity, antitumor inflammatory responses, and site-specific, enhanced drug/gene delivery. The application of mild hyperthermia increases blood flow and enhanced vascular permeability for effective delivery of chemotherapeutic agents. In addition, mild hyperthermia may serve to induce antitumor effects upon heat activation of gene products after successful delivery of a gene therapeutic.

Through in vivo delivery of the gene encoding heat shock protein (HSP) HSP70 into B16 melanoma nodules, Ito et al. (2003) have observed upregulated expression of HSP70 that led to antitumor effects after induced hyperthermia. Significant tumor arrest contributed to elevated cytotoxic T-lymphocyte (CTL) response against B16 melanoma in conjunction with prolonged survival/complete tumor regression in three of the ten mice. The combination of HSP70 gene therapy and hyperthermia could be an effective cancer therapeutic, as HSP70 may induce innate immunity by acting as a proinflammatory cytokine and induce adaptive immunity by T cell activation (Ito et al. 2003).

In addition to cancer therapy, the favorable effects of hyperthermia can be applied for gene delivery. Hyperthermia-induced downregulation of VE-cadherin is thought to contribute to rapid and reversible changes in endothelial cell permeability that can therefore enhance extravasation and promote drug/gene delivery (Böhmer et al. 2009; Yudina and Moonen 2012). Hyperthermia can initiate controlled release when used in conjunction with temperature-sensitive drug carriers. Yudina et al. (2012) demonstrated the successful delivery of cell-impermeable compounds in a two-step approach combining sonoporation and hyperthermic controlled release (Fig. 2.20). Encapsulation of cell-impermeable TO-PRO-3 fluorescent dye in a temperature-sensitive liposome (TSL) increased the dye's half-life and protected it from plasma degradation. Using a two-step mechanism, site-specific delivery was achieved by hyperthermic destabilization of TSL followed by ultrasound-mediated delivery through sonoporation (Fig. 2.20) (Yudina et al. 2012). As ultrasound-induced permeability lasts for prolonged periods, it is feasible to induce sonoporation prior to hyperthermic dye release; this was confirmed in a separate study by Yudina et al. (Yudina et al. 2011) where in vivo administration of TSLs to mice induced a 2.4-fold increase in fluorescence through sequential applications of sonoporation and hyperthermic dye release. Such results hold promise for hyperthermia-induced gene delivery.

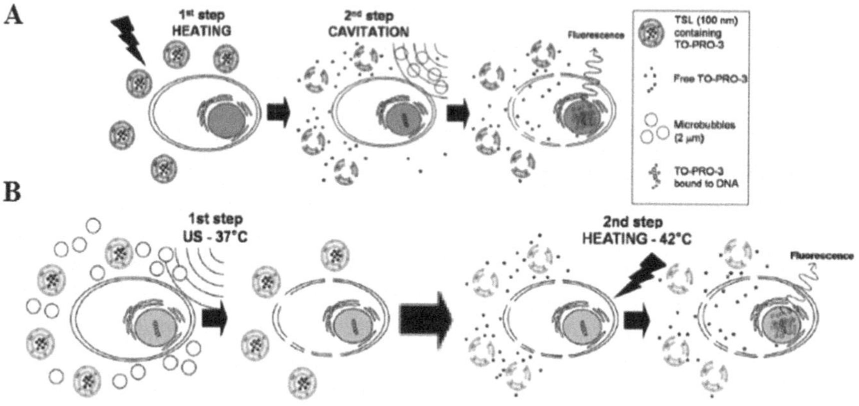

Fig. 2.20 Two-step mechanism for delivery of cell-impermeable TO-PRO-3: (**a**) hyperthermic release followed by sonoporation or (**b**) sonoporation followed by hyperthermic release. Adapted from Yudina et al. (2012, 2011)

Hyperthermic conditions pertinent to temperature-sensitive delivery can be achieved by radiofrequency, thermal effects of ultrasound, magnetic particles, and infrared light (Böhmer et al. 2009; Walther and Stein 2009). Regardless of the technique, the heat source must stably reach the desired temperatures in a temporally and spatially controlled manner for activation of temperature-sensitive carriers (Kim et al. 2012). Focused ultrasound and magnetic nanoparticles are commonly employed for temporal and spatial control of heat-responsive gene therapy. Focused ultrasound is generally combined with MRI for real-time monitoring and feedback control of applied temperatures (Böhmer et al. 2009; Sanches et al. 2011; Yudina and Moonen 2012; Walther and Stein 2009). Magnetic nanoparticles, incorporated into temperature-sensitive carriers, are employed for magnetically induced hyperthermia by an alternating magnetic field; these magnetic nanoparticles continuously emit heat through the Neel and Brownian relaxation pathways (Yoo et al. 2011). Focused ultrasound and magnetic nanoparticles are effective means of inducing hyperthermia, but they must be closely regulated such that hyperthermia can be induced without damaging surrounding tissues.

Temperature-sensitive delivery is achieved through the administration of TSLs or polymers. These liposomes/polymers must be optimized such that effective encapsulation can occur at physiological temperatures while achieving rapid release under hyperthermic conditions. TSLs are liposomes comprised of lipids that exhibit a gel-to-liquid crystalline phase transition at hyperthermic temperatures between 40 and 45 °C; phase transition causes transient pore formation that leads to leaky bilayer and rapid release (Böhmer et al. 2009; Sanches et al. 2011; Shao et al. 2011). Newer formulations, comprised of lysolipids and PEGylated lipids, allow for enhanced circulation and stability at physiological temperatures while exhibiting hyperthermia-induced rapid release. PEGylated lipids provide TSLs with better pharmacokinetic properties that enhance stability and circulation times in vivo, while lysolipids promote transient pore formation for rapid release (Sanches et al. 2011). TSLs are generally employed for effective drug delivery, but they have the potential for gene delivery. Temperature-sensitive polymers (TSP) are polymers that undergo phase transition at temperatures above their lower critical solution temperature (LCST). Below LCST, these polymers are good solvents and stably assemble into micelles. At temperatures above LCST, the polymers undergo changes in conformation, solubility, and hydrophobic/hydrophilic balance; these changes lead to collapse and aggregation of polymers that result in enhanced gene delivery (Böhmer et al. 2009; Shao et al. 2011). In a study by Schwerdt et al. (2008), DNA polyplexes were constructed using PEI covalently bound to temperature-sensitive vinylpyrrolidone and poly(*N*-isopropylacrylamide) copolymer, and they elicited a 100-fold increase in transfection efficiencies in vitro and a 10-fold increased DNA deposition in vivo under hyperthermic conditions. The increased DNA deposition constitutes significant selective transgene expression. The enhanced transfection efficiencies are the result of phase transition of the copolymers causing polymer aggregation and accumulation of polyplexes at target site. TSPs could be attached to liposomes for the generation of thermosensitive polymer-modified liposomes; such formulations possess improved stability and release kinetics through liposomal destabilization as a result of polymer contraction (Shao et al. 2011). Kaiden et al. (2011) generated a pH and temperature-sensitive polymer-modified lipoplex that destabilized and released its encapsulated nucleic acid cargo under mildly hyperthermic and mildly acidic conditions (Fig. 2.21). Synthesis of these thermosensitive polymer-modified liposomes involved the attachment of hyperbranched poly (glycidols), possessing temperature-sensitive *N*-isopropylamide and pH-sensitive succinylate (Suc) groups, to stabilize the egg yolk phosphatidylcholine liposomes. The dual-responsive lipoplexes could then undergo heat-induced phase transitions under mildly acidic conditions within endosomes; this resulted in destabilization of

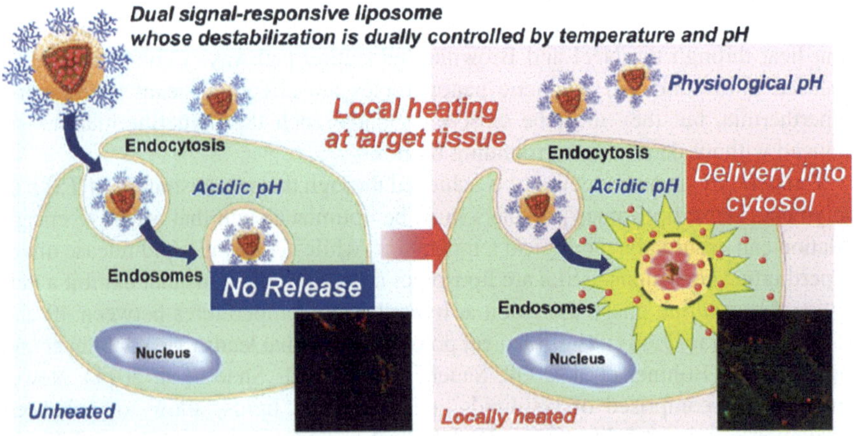

Fig. 2.21 Effective endosomal escape by pH- and temperature-sensitive, dual signal-responsive liposomes. Adapted from Kaiden et al. (2011)

the acidic endosome and subsequent release of nucleic acid into the cytosol. Undesirable release into extracellular space was avoided as temperature-induced destabilization was confined within endosomes after endocytosis (Fig. 2.21) (Kaiden et al. 2011). Hence, this has the potential of overcoming endosomal escape, a significant barrier to gene delivery. TSPs exhibit high versatility as their LCST can be altered depending on the ratio of hydrophilic and hydrophobic monomers in its composition (Shao et al. 2011); however, the phase transition of TSPs is difficult to control as LCST differs under in vitro and in vivo settings (Böhmer et al. 2009).

In addition to delivery, hyperthermia has been employed to induce transgene expression through the use of heat-responsive promoters. Heat-induced expression is mediated by heat-inducible promoters possessing heat shock elements and the subsequent binding of heat shock factors to those elements (Walther and Stein 2009). The use of endogenous pathways and transcription factors constitutes a one-component inducible system that overcomes several limitations (use of modified transcription factors and slow diffusion rate of inducers) that are present in two-component systems (Ortner et al. 2012). The magnitude of induced expression is dependent on temperature, duration, and cell type. Controllable transgene expression by heat-responsive promoters and hyperthermia can promote synergistic effect as seen in combinatorial cancer therapy through thermal ablation and induced transgene expression of suicide genes/cytokines. However, it is critical to define the correlation between temperature increase and expression such that transgene expression can be effectively controlled. Also, unintentional activation of heat-responsive promoters must be prevented so that induced expression can be deemed safe for implementation in gene therapy.

Heat shock protein 70B (HSP70B), growth arrest and DNA damage 153 gene (GADD153), human multidrug resistance gene 1 (MDR1/ABCB1), and cytomegalovirus immediate early (CMV-IE) are four heat-inducible promoters used in research and clinical applications, with HSP70B promoters as the most commonly used due to its low leakiness and high level of heat-induced transgene expression (Walther and Stein 2009). The addition of consensus heat shock factor protein 1 (HSF-1) binding sequences can optimize induced expression, while the use of a minimal HSP70B cassette ensures heat-specific activation as wild-type HSP70B are responsive to both heat and stress activation (Walther and Stein 2009). HSP70B has been used for induced suicide or cytokine (TNFα or IL-12) expression in cancer therapeutics. Tang et al. observed a 500-fold increase in transgene expression in vitro and higher concentration of gene product after 24 h in vivo from a combination of HSP70B and hyperthermia (Tang et al. 2008). GADD153 promoters are inducible by a variety of stressors including reactive oxygen species (ROS), DNA damage, cytotoxic drug effects, and heat. Through GADD153 promoters, Ito et al. (Ito et al. 2001) demonstrated that combinatorial therapy of hyperthermic ablation and gene delivery has promise as a potential viable cancer therapy; a threefold increase in TNFα was induced in the peripheral areas of a heated tumor environment where hyperthermic conditions were insufficient for direct cell death. Similar to GADD153, MDR1/ABCB1 promoters are stress-inducible promoters possessing heat shock elements that mediate heat-induced expression. CMV-IE promoters have been previously suggested to elevate transgene expression upon heat induction despite the fact that they do not possess any heat shock elements (Walther and Stein 2009). Through CMV-IE promoter activation, Kobelt et al. (2010) have shown the upregulated gene expression corresponding to a 25-fold increase in vitro and 2.5-fold in vivo. The effects of heat-induced activation of CMV-IE were presumed to be due to the activation of specific cellular pathways and transcription factors at a later time after induced hyperthermia (Kobelt et al. 2010).

The use of HSP70 promoter for controlled gene expression was further refined using MRI-guided high-intensity focused ultrasound (HIFU) for noninvasive controlled gene delivery. Deckers et al. (2009) demonstrated this spatiotemporal control of gene delivery and expression in a transgenic mouse model, observing transient gene expression 6–8 h after short thermal induction (approximately 2 min) without significant tissue damage. Increased duration (8 min) of hyperthermia did result in muscle damage, but such long duration was not necessary for gene delivery (Deckers et al. 2009). They later demonstrated this system again (Eker et al. 2011), whereby the gene encoding luciferase (*lucF*) under control of HSP70B was transfected into C6 cells, which were then injected and implanted into rat kidneys. Gene expression was then activated noninvasively using focused ultrasound, guided using MRI. Again, transient in vivo luciferase expression was observed 6–8 h after heat induction. More recently, the group turned to magneto-activatable thermogenic nanoparticles (MTN) (Sandre et al. 2017), which would allow for magnetic induction of hyperthermia as opposed to ultrasound. This could permit deeper penetration into tissues, such as poorly accessible tumors. MTNs were tested upon application to murine skin or were injected into tumor space for hyperthermia induction; however, in this study, activation of the HSP70 promoter was not stable, or consistent temperature increase was not detectable. This intriguing and innovative combination of technologies has many attractive assets but will require more investigation. Overall, the use of focused ultrasound for spatially targeted hyperthermia serves as a highly useful tool for controlled gene therapy.

HSP70B promoters can also be combined with a modified tetracycline (Tet) controlled expression system to generate a one-component, heat-inducible transgene expression system possessing a positive feedback loop for transcriptional amplification. This modified system, developed by Yamaguichi et al. (2012), is highly suited for hyperthermia-induced gene therapy as it merges the benefits of a tightly controlled, one-component heat-inducible system in HSP70B with transcriptional amplification from the Tet-OFF system (Figs. 2.22 and 2.23). In this system, heat activation of HSP70B promoter drives the initial expression of the Tet-responsive transcriptional activators (tTA); the subsequent binding of tTA to the TRE/PCMVmin promoter, situated downstream of the HSP70B promoter, constitutes a positive feedback loop for amplification of tTA expression. tTA serves to activate and amplify therapeutic gene expression downstream of a TRE/PCMVmin promoter (see Fig. 2.22) (Yamaguchi et al. 2012). Using this system, a 3000- and 6500-fold increase in expression was observed 48 h after heat induction at 43 °C and 45 °C, respectively; however, this required the successful co-transfection of both plasmids into the same cell in order for inducible expression to arise. To simplify the

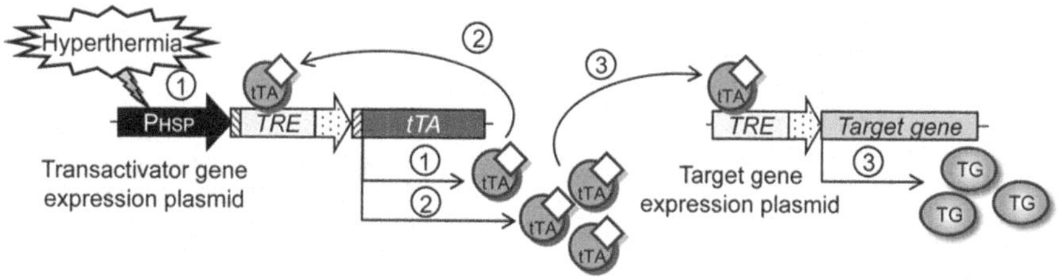

Fig. 2.22 Schematic illustration of a two-plasmid heat-inducible gene expression system. Adapted from Yamaguchi et al. (2012)

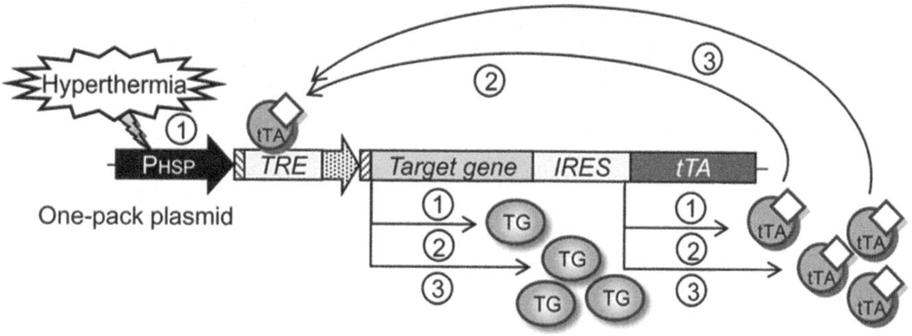

Fig. 2.23 Schematic illustration of a one-pack plasmid heat-inducible gene expression system. Adapted from Yamaguchi et al. (2012)

system, a one-pack plasmid was generated such that the expression of both tTA and therapeutic gene could be simultaneously amplified (see Fig. 2.23) (Yamaguchi et al. 2012). The one-pack plasmid was able to yield expression levels higher than the two-plasmid approach; this was mainly attributed to higher transfection efficiencies. The enhanced expression of the one-pack plasmid was confirmed to be due to the positive feedback loop as the addition of doxycycline, a tetracycline analog capable of inhibiting expression in traditional Tet-OFF systems, resulting in a sixfold reduction in expression (Yamaguchi et al. 2012). Findings from this study have serious in vivo implications as high transgene expression can occur at low transfection efficiencies. The authors have successfully applied the system for suicide gene therapy using herpes simplex virus thymidine kinase gene (HSV-tk) as the therapeutic gene; the high levels of cytotoxic effect observed is significantly greater than cytotoxic effects due to heat-induced apoptosis. Thus, the potential for effective combinatorial cancer therapy has been demonstrated.

The application of natural heat-inducible promoters for expression upregulation may be problematic as they are commonly integrated into other emergency pathways not mediated by heat shock factors (HSF-1), which could contribute to tissue-specific variations (Ortner et al. 2012). Such limitations can be bypassed through the generation of artificial heat-inducible promoters. In a study by Ortner et al. (2012), stable transfection and subsequent heat-induced gene expression was

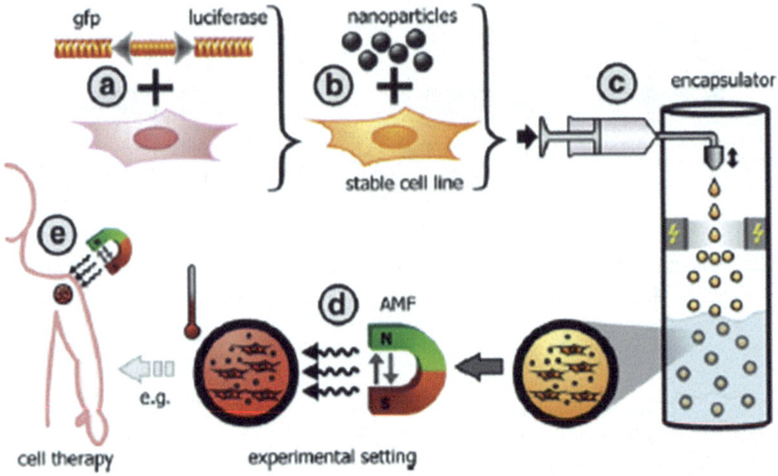

Fig. 2.24 Schematic illustrating the application of encapsulation systems in cell therapy. Adapted from Ortner et al. (2012)

observed through the application of an artificial heat shock-inducible promoter consisting of two minimal CMV promoter elements flanking multiple high-affinity heat shock elements (Fig. 2.24). The artificial promoter elicited similar kinetics to natural heat shock promoters, and heat-induced expressions were highly regulated depending on the duration of induced hyperthermia. The use of magnetic nanoparticles and alternating magnetic field for heat generation allows for efficient modulation of gene expression. The authors were able to demonstrate the encapsulation of magnetic nanoparticles and stably transfected HEK293, possessing the aforementioned artificial promoter, into sodium cellulose sulfate (SCS) capsules. The encapsulation of SCS capsules had no effect on cell viability as cells continued to divide within the capsule. Exposure to alternating magnetic field resulted in 1700- and 950-fold increases in luciferase and GFP gene expression in vitro (Sanford et al. 1987). Such results signify the potential of encapsulation systems for cell therapy as they possess several advantages: (1) protection of genetically modified cells from immune activation, (2) cell selection and testing prior to encapsulation for optimized therapeutic gene expression, (3) inducible expression of therapeutic genes, and (4) remote control of induced expression by the application of alternating magnetic field (Ortner et al. 2012).

References

Ajima K, Murakami T, Mizoguchi Y et al (2008) Enhancement of in vivo anticancer effects of cisplatin by incorporation inside single-wall carbon nanohorns. ACS Nano 2:2057–2064. https://doi.org/10.1021/nn800395t

Al-Deen FN, Ho J, Selomulya C et al (2011) Superparamagnetic nanoparticles for effective delivery of malaria DNA vaccine. Langmuir 27:3703–3712. https://doi.org/10.1021/la104479c

Al-Dosari MS, Gao X (2009) Nonviral gene delivery: principle, limitations, and recent progress. AAPS J 11:671–681. https://doi.org/10.1208/s12248-009-9143-y

Allen TM (1994) Long-circulating (sterically stabilized) liposomes for targeted drug delivery. Trends Pharmacol Sci 15:215–220. https://doi.org/10.1016/0165-6147(94)90314-X

Alzaraa A, Gravante G, Chung WY et al (2012) Targeted microbubbles in the experimental and clinical setting. Am J Surg 204:355–366. https://doi.org/10.1016/j.amjsurg.2011.10.024

Antkowiak M, Torres-Mapa ML, Gunn-Moore F, Dholakia K (2010) Application of dynamic diffractive optics for enhanced femtosecond laser based cell transfection. J Biophoton 3:696–705. https://doi.org/10.1002/jbio.201000052

Arsianti M, Lim M, Marquis CP, Amal R (2010) Polyethylenimine based magnetic iron-oxide vector: the effect of vector component assembly on cellular entry mechanism, intracellular localization, and cellular viability. Biomacromolecules 11:2521–2531. https://doi.org/10.1021/bm100748p

Bakry R, Vallant RM, Najam-ul-Haq M et al (2007) Medicinal applications of fullerenes. Int J Nanomed 2:639–649

Bandow S, Kokai F, Takahashi K et al (2000) Interlayer spacing anomaly of single-wall carbon nanohorn aggregate. Chem Phys Lett 321:514–519. https://doi.org/10.1016/S0009-2614(00)00353-5

Barhoumi A, Huschka R, Bardhan R et al (2009) Light-induced release of DNA from plasmon-resonant nanoparticles: towards light-controlled gene therapy. Chem Phys Lett 482:171–179. https://doi.org/10.1016/j.cplett.2009.09.076

Baumgart J, Humbert L, Boulais É et al (2012) Off-resonance plasmonic enhanced femtosecond laser optoporation and transfection of cancer cells. Biomaterials 33:2345–2350. https://doi.org/10.1016/j.biomaterials.2011.11.062

Behr J-P (1997) The proton sponge: a trick to enter cells the viruses did not exploit. Int J Chem 2:34–36

Böhmer MR, Klibanov AL, Tiemann K et al (2009) Ultrasound triggered image-guided drug delivery. Eur J Radiol 70:242–253. https://doi.org/10.1016/j.ejrad.2009.01.051

Boisselier E, Astruc D (2009) Gold nanoparticles in nanomedicine: preparations, imaging, diagnostics, therapies and toxicity. Chem Soc Rev 38:1759–1782. https://doi.org/10.1039/b806051g

Boussif O, Lezoualc'h F, Zanta MA et al (1995) A versatile vector for gene and oligonucleotide transfer into cells in culture and in vivo: polyethylenimine. Proc Natl Acad Sci U S A 92:7297–7301

Braun GB, Pallaoro A, Wu G et al (2009) Laser-activated gene silencing via gold nanoshell-siRNA conjugates. ACS Nano 3:2007–2015. https://doi.org/10.1021/nn900469q

Cai D, Mataraza JM, Qin Z-H et al (2005) Highly efficient molecular delivery into mammalian cells using carbon nanotube spearing. Nat Methods 2:449–454. https://doi.org/10.1038/nmeth761

Cebrián V, Martín-Saavedra F, Yagüe C et al (2011) Size-dependent transfection efficiency of PEI-coated gold nanoparticles. Acta Biomater 7:3645–3655. https://doi.org/10.1016/j.actbio.2011.06.018

Chen Z, Liang K, Liu J et al (2009a) Enhancement of survivin gene downregulation and cell apoptosis by a novel combination: liposome microbubbles and ultrasound exposure. Med Oncol 26:491–500. https://doi.org/10.1007/s12032-008-9161-0

Chen Z, Liang K, Xie M et al (2009b) Novel ultrasound-targeted microbubble destruction mediated short hairpin RNA plasmid transfection targeting survivin inhibits gene expression and induces apoptosis of HeLa cells. Mol Biol Rep 36:2059–2067. https://doi.org/10.1007/s11033-008-9417-y

Chen Z-Y, Liang K, Qiu R-X (2010) Targeted gene delivery in tumor xenografts by the combination of ultrasound-targeted microbubble destruction and polyethylenimine to inhibit survivin gene expression and induce apoptosis. J Exp Clin Cancer Res 29:152. https://doi.org/10.1186/1756-9966-29-152

Cheung W, Pontoriero F, Taratula O et al (2010) DNA and carbon nanotubes as medicine. Adv Drug Deliv Rev 62:633–649. https://doi.org/10.1016/j.addr.2010.03.007

Chonn A, Semple SC, Cullis PR (1992) Association of blood proteins with large unilamellar liposomes in vivo. J Biol Chem 267:18759–18765

Cizmár T, Kollárová V, Tsampoula X et al (2008) Generation of multiple Bessel beams for a biophotonics workstation. Opt Express 16:14024–14035

Cline MJ (1985) Perspectives for gene therapy: inserting new genetic information into mammalian cells by physical techniques and viral vectors. Pharmacol Ther 29:69–92

Conde J, Ambrosone A, Sanz V et al (2012) Design of multifunctional gold nanoparticles for in vitro and in vivo gene silencing. ACS Nano 6:8316–8324. https://doi.org/10.1021/nn3030223

Cool SK, Geers B, Lentacker I et al (2013) Enhancing nucleic acid delivery with ultrasound and microbubbles. Methods Mol Biol 948:195–204

Cui D, Huang P, Zhang C et al (2011) Dendrimer-modified gold nanorods as efficient controlled gene delivery system under near-infrared light irradiation. J Control Release 152:e137–e139. https://doi.org/10.1016/j.jconrel.2011.08.047

Cutler JI, Auyeung E, Mirkin CA (2012) Spherical nucleic acids. J Am Chem Soc 134:1376–1391. https://doi.org/10.1021/ja209351u

Deckers R, Quesson B, Arsaut J et al (2009) Image-guided, noninvasive, spatiotemporal control of gene expression. Proc Natl Acad Sci U S A 106:1175–1180. https://doi.org/10.1073/pnas.0806936106

Delalande A, Kotopoulis S, Rovers T et al (2011) Sonoporation at a low mechanical index. Bubble Sci Eng Technol 3:3–12. https://doi.org/10.1179/1758897911Y.0000000001

Delalande A, Postema M, Mignet N et al (2012) Ultrasound and microbubble-assisted gene delivery: recent advances and ongoing challenges. Ther Deliv 3:1199–1215

Dholakia K (2010) Phototransfection of mammalian cells using femtosecond laser pulses: optimization and applicability to stem cell differentiation. J Biomed Opt 15:1–7. https://doi.org/10.1117/1.3430733

Donate A, Bulysheva A, Edelblute C et al (2016) Thermal assisted in vivo gene electrotransfer. Curr Gene Ther 16:83–89

Eker OF, Quesson B, Rome C et al (2011) Combination of cell delivery and thermoinducible transcription for in vivo spatiotemporal control of gene expression: a feasibility study. Radiology 258:496–504. https://doi.org/10.1148/radiol.10100767

Endo-Takahashi Y, Negishi Y, Kato Y et al (2012) Efficient siRNA delivery using novel siRNA-loaded bubble liposomes and ultrasound. Int J Pharm 422:504–509. https://doi.org/10.1016/j.ijpharm.2011.11.023

Ensenauer R, Hartl D, Vockley J et al (2011) Efficient and gentle siRNA delivery by magnetofection. Biotech Histochem 86:226–231. https://doi.org/10.3109/10520291003675485

Escriou V, Carrière M, Scherman D, Wils P (2003) NLS bioconjugates for targeting therapeutic genes to the nucleus. Adv Drug Deliv Rev 55:295–306. https://doi.org/10.1016/S0169-409X(02)00184-9

Ewert K, Slack NL, Ahmad A et al (2004) Cationic lipid-DNA complexes for gene therapy: understanding the relationship between complex structure and gene delivery pathways at the molecular level. Curr Med Chem 11:133–149

Fallini C, Bassell GJ, Rossoll W (2010) High-efficiency transfection of cultured primary motor neurons to study protein localization, trafficking, and function. Mol Neurodegener 5:17. https://doi.org/10.1186/1750-1326-5-17

Fan X, Tan J, Zhang G, Zhang F (2007) Isolation of carbon nanohorn assemblies and their potential for intracellular delivery. Nanotechnology 18:195103. https://doi.org/10.1088/0957-4484/18/19/195103

Farhood H, Serbina N, Huang L (1995) The role of dioleoyl phosphatidylethanolamine in cationic liposome mediated gene transfer. Biochim Biophys Acta Biomembr 1235:289–295. https://doi.org/10.1016/0005-2736(95)80016-9

Felgner PL, Gadek TR, Holm M et al (1987) Lipofection: a highly efficient, lipid-mediated DNA-transfection procedure. Proc Natl Acad Sci U S A 84:7413–7417. https://doi.org/10.1073/pnas.84.21.7413

Feng L, Zhang S, Liu Z (2011) Graphene based gene transfection. Nanoscale 3:1252–1257. https://doi.org/10.1039/c0nr00680g

Feng L, Yang X, Shi X et al (2013) Polyethylene glycol and polyethylenimine dual-functionalized nano-graphene oxide for photothermally enhanced gene delivery. Small 9:1989–1997. https://doi.org/10.1002/smll.201202538

Feramisco J, Perona R, Lacal JC (1999) Needle microinjection: a brief history. In: Microinjection. Birkhäuser Basel, Basel, pp 9–15

Feril LB, Tachibana K (2012) Use of ultrasound in drug delivery systems: emphasis on experimental methodology and mechanisms. Int J Hyperthermia 28:282–289. https://doi.org/10.3109/02656736.2012.668640

Ferrari AC, Bonaccorso F, Fal'ko V et al (2015) Science and technology roadmap for graphene, related two-dimensional crystals, and hybrid systems. Nanoscale 7:4598–4810. https://doi.org/10.1039/C4NR01600A

Figueiredo M, Esenaliev R (2012) PLGA nanoparticles for ultrasound-mediated gene delivery to solid tumors. J Drug Deliv 2012:1–20. https://doi.org/10.1155/2012/767839

Fraley R, Subramani S, Berg P, Papahadjopoulos D (1980) Introduction of liposome-encapsulated SV40 DNA into cells. J Biol Chem 255:10431–10435

Gebeyehu G, Jessee JA, Ciccarone VC, Hawley-Nelson P, Chytil A (1992) U.S. Patent No. US5334761A. U.S. Patent and Trademark Office, Washington, DC

Gehl J (2014) Gene electrotransfer in clinical trials. In: Li S (ed) Electroporation protocols: preclinical and clincial gene medicine. Springer, Berlin, pp 241–246

Geis NA, Katus HA, Bekeredjian R (2012) Microbubbles as a vehicle for gene and drug delivery: current clinical implications and future perspectives. Curr Pharm Des 18:2166–2183

Ghosh P, Han G, De M et al (2008) Gold nanoparticles in delivery applications. Adv Drug Deliv Rev 60:1307–1315. https://doi.org/10.1016/j.addr.2008.03.016

Ghosh R, Singh LC, Shohet JM, Gunaratne PH (2013) A gold nanoparticle platform for the delivery of functional microRNAs into cancer cells. Biomaterials 34:807–816. https://doi.org/10.1016/j.biomaterials.2012.10.023

Gothelf A, Gehl J (2010) Gene electrotransfer to skin; Review of existing literature and clinical perspectives. Curr Gene Ther 10:287–299. https://doi.org/10.2174/156652310791823443

Guerra J, Herrero MA, Vázquez E (2014) Carbon nanohorns as alternative gene delivery vectors. RSC Adv 4:27315. https://doi.org/10.1039/c4ra03251a

Gul-Uludag H, Lu W, Xu P et al (2012) Efficient and rapid uptake of magnetic carbon nanotubes into human monocytic cells: implications for cell-based cancer gene therapy. Biotechnol Lett 34:989–993. https://doi.org/10.1007/s10529-012-0858-y

Guo C, Al-Jamal WT, Toma FM et al (2015) Design of cationic multiwalled carbon nanotubes as efficient siRNA vectors for lung cancer xenograft eradication. Bioconjug Chem 26:1370–1379. https://doi.org/10.1021/acs.bioconjchem.5b00249

Hacein-Bey Abina S, von Kalle C, Schmidt M et al (2003) A serious adverse event after successful gene therapy for X-linked severe combined immunodeficiency. N Engl J Med 348:255–266. https://doi.org/10.1056/NEJM200301163480314

Hacein-bey-abina S, Garrigue A, Wang GP et al (2008) Insertional oncogenesis in 4 patients after retrovirus-mediated gene therapy of SCID-X1. J Clin Invest 118:3132–3142. https://doi.org/10.1172/JCI35700.3132

Harasym TO, Tardi P, Longman SA et al (1995) Poly(ethylene glycol)-modified phospholipids prevent aggregation during covalent conjugation of proteins to liposomes. Bioconjug Chem 6:187–194. https://doi.org/10.1021/bc00032a006

He H, Kong SK, Chan KT (2011) Transfection and cell fusion by femtosecond lasers. J Innov Opt Health Sci 4:113–125. https://doi.org/10.1142/S179354581100123X

Heller R, Heller LC (2015) Gene electrotransfer clinical trials. Adv Genet 89:235–262. https://doi.org/10.1016/bs.adgen.2014.10.006

Horie S, Watanabe Y, Chen R et al (2010) Development of localized gene delivery using a dual-intensity ultrasound system in the bladder. Ultrasound Med Biol 36:1867–1875. https://doi.org/10.1016/j.ultrasmedbio.2010.07.015

Horie S, Watanabe Y, Ono M et al (2011) Evaluation of antitumor effects following tumor necrosis factor-α gene delivery using nanobubbles and ultrasound. Cancer Sci 102:2082–2089. https://doi.org/10.1111/j.1349-7006.2011.02056.x

Huang S-L (2008) Liposomes in ultrasonic drug and gene delivery. Adv Drug Deliv Rev 60:1167–1176. https://doi.org/10.1016/j.addr.2008.03.003

Huang H-C, Barua S, Kay DB, Rege K (2009) Simultaneous enhancement of photothermal stability and gene delivery efficacy of gold nanorods using polyelectrolytes. ACS Nano 3:2941–2952. https://doi.org/10.1021/nn900947a

Hui SW, Langner M, Zhao YL et al (1996) The role of helper lipids in cationic liposome-mediated gene transfer. Biophys J 71:590–599. https://doi.org/10.1016/S0006-3495(96)79309-8

Huschka R, Neumann O, Barhoumi A, Halas NJ (2010) Visualizing light-triggered release of molecules inside living cells. Nano Lett 10:4117–4122. https://doi.org/10.1021/nl102293b

Huschka R, Zuloaga J, Knight MW et al (2011) Light-induced release of DNA from gold nanoparticles: nanoshells and nanorods. J Am Chem Soc 133:12247–12255. https://doi.org/10.1021/ja204578e

Husseini GA, Pitt WG (2008) Micelles and nanoparticles for ultrasonic drug and gene delivery. Adv Drug Deliv Rev 60:1137–1152. https://doi.org/10.1016/j.addr.2008.03.008

Iijima S, Yudasaka M, Yamada R et al (1999) Nano-aggregates of single-walled graphitic carbon nano-horns. Chem Phys Lett 309:165–170. https://doi.org/10.1016/S0009-2614(99)00642-9

Immordino ML, Dosio F, Cattel L et al (2006) Stealth liposomes: review of the basic science, rationale, and clinical applications, existing and potential. Int J Nanomed 1:297–315. https://doi.org/10.1023/A:1020134521778

Ishii A, Hiruta Y, Heinemann D et al (2017) Intracellular localization and delivery of plasmid DNA by biodegradable microsphere-mediated femtosecond laser optoporation. J Biophoton 1–10. https://doi.org/10.1002/jbio.201600323

Isobe H, Nakanishi W, Tomita N et al (2006a) Nonviral gene delivery by tetraamino fullerene. Mol Pharm 3:124–134. https://doi.org/10.1021/mp050068r

Isobe H, Nakanishi W, Tomita N et al (2006b) Gene delivery by aminofullerenes: structural requirements for efficient transfection. Chem Asian J 1:167–175. https://doi.org/10.1002/asia.200600051

Ito A, Shinkai M, Honda H, Kobayashi T (2001) Heat-inducible TNF-α gene therapy combined with hyperthermia using magnetic nanoparticles as a novel tumor-targeted therapy. Cancer Gene Ther 8:649–654. https://doi.org/10.1038/sj.cgt.7700357

Ito A, Matsuoka F, Honda H, Kobayashi T (2003) Heat shock protein 70 gene therapy combined with hyperthermia using magnetic nanoparticles. Cancer Gene Ther 10:918–925. https://doi.org/10.1038/sj.cgt.7700648

Jenkins SI, Pickard MR, Granger N, Chari DM (2011) Magnetic nanoparticle-mediated gene transfer to oligodendrocyte precursor cell transplant populations is enhanced by magnetofection strategies. ACS Nano 5:6527–6538. https://doi.org/10.1021/nn2018717

Jin H, Heller DA, Sharma R, Strano MS (2009) Size-dependent cellular uptake and expulsion of single-walled carbon nanotubes: single particle tracking and a generic uptake model for nanoparticles. ACS Nano 3:149–158. https://doi.org/10.1021/nn800532m

Kadlecova Z, Rajendra Y, Matasci M et al (2013) DNA delivery with hyperbranched polylysine: a comparative study with linear and dendritic polylysine. J Control Release 169:276–288. https://doi.org/10.1016/j.jconrel.2013.01.019

Kaiden T, Yuba E, Harada A et al (2011) Dual signal-responsive liposomes for temperature-controlled cytoplasmic delivery. Bioconjug Chem 22:1909–1915. https://doi.org/10.1021/bc2000353

Kam NWS, Dai H (2005) Carbon nanotubes as intracellular protein transporters: generality and biological functionality. J Am Chem Soc 127:6021–6026. https://doi.org/10.1021/JA050062V

Kam NWS, Liu Z, Dai H (2006) Carbon nanotubes as intracellular transporters for proteins and DNA: an investigation of the uptake mechanism and pathway. Angew Chem Int Ed 45:577–581. https://doi.org/10.1002/anie.200503389

Kamau SW, Hassa PO, Steitz B et al (2006) Enhancement of the efficiency of non-viral gene delivery by application of pulsed magnetic field. Nucleic Acids Res 34:e40. https://doi.org/10.1093/nar/gkl035

Kami D, Takeda S, Itakura Y et al (2011) Application of magnetic nanoparticles to gene delivery. Int J Mol Sci 12:3705–3722. https://doi.org/10.3390/ijms12063705

Kang B, Chang S, Dai Y et al (2010) Cell response to carbon nanotubes: size-dependent intracellular uptake mechanism and subcellular fate. Small 6:2362–2366. https://doi.org/10.1002/smll.201001260

Keswani RK, Lazebnik M, Pack DW (2015) Intracellular trafficking of hybrid gene delivery vectors. J Control Release 207:120–130. https://doi.org/10.1016/j.jconrel.2015.04.015

Kiessling F, Fokong S, Koczera P et al (2012) Ultrasound microbubbles for molecular diagnosis, therapy, and theranostics. J Nucl Med 53:345–348. https://doi.org/10.2967/jnumed.111.099754

Kim J, Hwang I, Britain D et al (2011) Microfluidic approaches for gene delivery and gene therapy. Lab Chip 11:3941–3948. https://doi.org/10.1039/c1lc20766k

Kim Y-C, Park J-H, Prausnitz MR (2012) Microneedles for drug and vaccine delivery. Adv Drug Deliv Rev 64:1547–1568. https://doi.org/10.1016/j.addr.2012.04.005

Klumpp C, Lacerda L, Chaloin O et al (2007) Multifunctionalised cationic fullerene adducts for gene transfer: design, synthesis and DNA complexation. Chem Commun 66:3762–3764. https://doi.org/10.1039/b708435h

Kobelt D, Aumann J, Fichtner I et al (2010) Activation of the CMV-IE promoter by hyperthermia in vitro and in vivo: biphasic heat induction of cytosine deaminase suicide gene expression. Mol Biotechnol 46:197–205. https://doi.org/10.1007/s12033-010-9292-3

Kodama T, Aoi A, Watanabe Y et al (2010) Evaluation of transfection efficiency in skeletal muscle using nano/microbubbles and ultrasound. Ultrasound Med Biol 36:1196–1205. https://doi.org/10.1016/j.ultrasmedbio.2010.04.016

Kohli V, Robles V, Cancela ML et al (2007) An alternative method for delivering exogenous material into developing zebrafish embryos. Biotechnol Bioeng 98:1230–1241. https://doi.org/10.1002/bit.21564

Koltover I, Salditt T, Rädler JO, Safinya CR (1998) An inverted hexagonal phase of cationic liposome-DNA complexes related to DNA release and delivery. Science 281:78–81

Kong M, Li X, Wang C et al (2012a) Tissue distribution and cancer growth inhibition of magnetic lipoplex-delivered type 1 insulin-like growth factor receptor shRNA in nude mice. Acta Biochim Biophys Sin (Shanghai) 44:591–596. https://doi.org/10.1093/abbs/gms039

Kong WH, Bae KH, Jo SD et al (2012b) Cationic lipid-coated gold nanoparticles as efficient and non-cytotoxic intracellular siRNA delivery vehicles. Pharm Res 29:362–374. https://doi.org/10.1007/s11095-011-0554-y

Kostarelos K, Lacerda L, Pastorin G et al (2007) Cellular uptake of functionalized carbon nanotubes is independent of functional group and cell type. Nat Nanotechnol 2:108–113. https://doi.org/10.1038/nnano.2006.209

Krasovitski B, Frenkel V, Shoham S, Kimmel E (2011) Intramembrane cavitation as a unifying mechanism for ultrasound-induced bioeffects. Proc Natl Acad Sci U S A 108:3258–3263. https://doi.org/10.1073/pnas.1015771108

Kroto HW, Heath JR, O'Brien SC et al (1985) C60: Buckminsterfullerene. Nature 318:162–163. https://doi.org/10.1038/318162a0

Kurita A, Matsunobu T, Satoh Y et al (2011) Targeted gene transfer into rat facial muscles by nanosecond pulsed laser-induced stress waves. J Biomed Opt 16:98002. https://doi.org/10.1117/1.3628313

Lacerda L, Russier J, Pastorin G et al (2012) Translocation mechanisms of chemically functionalised carbon nanotubes across plasma membranes. Biomaterials 33:3334–3343. https://doi.org/10.1016/j.biomaterials.2012.01.024

Laemmli UK (1975) Characterization of DNA condensates induced by poly(ethylene oxide) and polylysine. Proc Natl Acad Sci U S A 72:4288–4292

Lam AP, Dean DA (2010) Progress and prospects: nuclear import of nonviral vectors. Gene Ther 17:439–447. https://doi.org/10.1038/gt.2010.31

Lee SE, Liu GL, Kim F, Lee LP (2009) Remote optical switch for localized and selective control of gene interference. Nano Lett 9:562–570. https://doi.org/10.1021/nl802689k

Lee SE, Sasaki DY, Park Y et al (2012) Photonic gene circuits by optically addressable siRNA-Au nanoantennas. ACS Nano 6:7770–7780. https://doi.org/10.1021/nn301744x

Lin AJ, Slack NL, Ahmad A et al (2003) Three-dimensional imaging of lipid gene-carriers: membrane charge density controls universal transfection behavior in lamellar cationic liposome-DNA complexes. Biophys J 84:3307–3316. https://doi.org/10.1016/S0006-3495(03)70055-1

Link S, El-Sayed MA (2003) Optical properties and ultrafast dynamics of metallic nanocrystals. Annu Rev Phys Chem 54:331–366. https://doi.org/10.1146/annurev.physchem.54.011002.103759

Liu W-M, Xue Y-N, He W-T et al (2011a) Dendrimer modified magnetic iron oxide nanoparticle/DNA/PEI ternary complexes: a novel strategy for magnetofection. J Control Release 152:e159–e160. https://doi.org/10.1016/j.jconrel.2011.08.061

Liu M, Chen B, Xue Y et al (2011b) Polyamidoamine-grafted multiwalled carbon nanotubes for gene delivery: synthesis, transfection and intracellular trafficking. Bioconjug Chem 22:2237–2243. https://doi.org/10.1021/bc200189f

Lu W, Zhang G, Zhang R et al (2010a) Tumor site-specific silencing of NF-kappaB p65 by targeted hollow gold nanosphere-mediated photothermal transfection. Cancer Res 70:3177–3188. https://doi.org/10.1158/0008-5472.CAN-09-3379

Lu C-H, Zhu C-L, Li J et al (2010b) Using graphene to protect DNA from cleavage during cellular delivery. Chem Commun 46:3116–3118. https://doi.org/10.1039/b926893f

Lukianova-Hleb EY, Wagner DS, Brenner MK, Lapotko DO (2012) Cell-specific transmembrane injection of molecular cargo with gold nanoparticle-generated transient plasmonic nanobubbles. Biomaterials 33:5441–5450. https://doi.org/10.1016/j.biomaterials.2012.03.077

Ma Y, Zhang Z, Wang X et al (2011) Insights into the mechanism of magnetofection using MNPs-PEI/pDNA/free PEI magnetofectins. Int J Pharm 419:247–254. https://doi.org/10.1016/j.ijpharm.2011.07.017

Maeda-Mamiya R, Noiri E, Isobe H et al (2010) In vivo gene delivery by cationic tetraamino fullerene. Proc Natl Acad Sci U S A 107:5339–5344. https://doi.org/10.1073/pnas.0909223107

Mallory M, Gogineni E, Jones GC et al (2016) Therapeutic hyperthermia: the old, the new, and the upcoming. Crit Rev Oncol Hematol 97:56–64. https://doi.org/10.1016/j.critrevonc.2015.08.003

Mannell H, Pircher J, Fochler F et al (2012) Site directed vascular gene delivery in vivo by ultrasonic destruction of magnetic nanoparticle coated microbubbles. Nanomedicine 8:1309–1318. https://doi.org/10.1016/j.nano.2012.03.007

Männistö M, Vanderkerken S, Toncheva V et al (2002) Structure–activity relationships of poly (l-lysines): effects of pegylation and molecular shape on physicochemical and biological properties in gene delivery. J Control Release 83:169–182. https://doi.org/10.1016/S0168-3659(02)00178-5

Mar Encabo-Berzosa M, Sancho-Albero M, Sebastian V et al (2017) Polymer functionalized gold nanoparticles as non-viral gene delivery reagents. J Gene Med:e2964. https://doi.org/10.1002/jgm.2964

Marchesan S, Kostarelos K, Bianco A, Prato M (2015) The winding road for carbon nanotubes in nanomedicine. Mater Today 18:12–19. https://doi.org/10.1016/j.mattod.2014.07.009

Matsui H, Johnson LG, Randell SH, Boucher RC (1997) Loss of binding and entry of liposome-DNA complexes decreases transfection efficiency in differentiated airway epithelial cells. J Biol Chem 272:1117–1126. https://doi.org/10.1074/jbc.272.2.1117

Matsushita-Ishiodori Y, Ohtsuki T (2012) Photoinduced RNA interference. Acc Chem Res 45:1039–1047. https://doi.org/10.1021/ar200227n

Matsushita-Ishiodori Y, Kuwabara R, Sakakoshi H et al (2011) Photosensitizing carrier proteins for photoinducible RNA interference. Bioconjug Chem 22:2222–2226. https://doi.org/10.1021/bc200095a

May S, Ben-Shaul A (2004) Modeling of cationic lipid-DNA complexes. Curr Med Chem 11:151–167

McBain SC, Griesenbach U, Xenariou S et al (2008) Magnetic nanoparticles as gene delivery agents: enhanced transfection in the presence of oscillating magnet arrays. Nanotechnology 19:405102. https://doi.org/10.1088/0957-4484/19/40/405102

Mead BP, Kim N, Miller GW et al (2017) Novel focused ultrasound gene therapy approach noninvasively restores dopaminergic neuron function in a rat Parkinson's disease model. Nano Lett 17:3533–3542. https://doi.org/10.1021/acs.nanolett.7b00616

Mellott AJ, Forrest ML, Detamore MS (2013) Physical non-viral gene delivery methods for tissue engineering. Ann Biomed Eng 41:446–468. https://doi.org/10.1007/s10439-012-0678-1

Meyer O, Kirpotin D, Hong K et al (1998) Cationic liposomes coated with polyethylene glycol as carriers for oligonucleotides. J Biol Chem 273:15621–15627. https://doi.org/10.1074/jbc.273.25.15621

Miao L, Zhang K, Qiao C et al (2013) Antitumor effect of human TRAIL on adenoid cystic carcinoma using magnetic nanoparticle–mediated gene expression. Nanomedicine 9:141–150. https://doi.org/10.1016/j.nano.2012.04.006

Miller AM, Dean DA (2009) Tissue-specific and transcription factor-mediated nuclear entry of DNA. Adv Drug Deliv Rev 61:603–613. https://doi.org/10.1016/j.addr.2009.02.008

Mintzer MA, Simanek EE (2009) Nonviral vectors for gene delivery. Chem Rev 109:259–302. https://doi.org/10.1021/cr800409e

Mishra S, Webster P, Davis ME (2004) PEGylation significantly affects cellular uptake and intracellular trafficking of non-viral gene delivery particles. Eur J Cell Biol 83:97–111. https://doi.org/10.1078/0171-9335-00363

Miyawaki J, Yudasaka M, Azami T et al (2008) Toxicity of single-walled carbon nanohorns. ACS Nano 2:213–226. https://doi.org/10.1021/nn700185t

Mochizuki S, Kanegae N, Nishina K et al (2013) The role of the helper lipid dioleoylphosphatidylethanolamine (DOPE) for DNA transfection cooperating with a cationic lipid bearing ethylenediamine. Biochim Biophys Acta Biomembr 1828:412–418. https://doi.org/10.1016/j.bbamem.2012.10.017

Moosavi Nejad S, Hosseini SHR, Akiyama H, Tachibana K (2011) Optical observation of cell sonoporation with low intensity ultrasound. Biochem Biophys Res Commun 413:218–223. https://doi.org/10.1016/j.bbrc.2011.08.072

Mulvana H, Eckersley RJ, Tang M-X et al (2012) Theoretical and experimental characterisation of magnetic microbubbles. Ultrasound Med Biol 38:864–875. https://doi.org/10.1016/j.ultrasmedbio.2012.01.027

Murakami T, Ajima K, Miyawaki J et al (2004) Drug-loaded carbon nanohorns: adsorption and release of dexamethasone in vitro. Mol Pharm 1:399–405. https://doi.org/10.1021/MP049928E

Murakami T, Fan J, Yudasaka M et al (2006) Solubilization of single-wall carbon nanohorns using a PEG-doxorubicin conjugate. Mol Pharm 3:407–414. https://doi.org/10.1021/mp060027a

Murata K, Kaneko K, Steele WA et al (2001) Molecular potential structures of heat-treated single-wall carbon nanohorn assemblies. J Phys Chem B 105:10210–10216. https://doi.org/10.1021/jp010754f

Mykhaylyk O, Antequera YS, Vlaskou D, Plank C (2007) Generation of magnetic nonviral gene transfer agents and magnetofection in vitro. Nat Protoc 2:2391–2411. https://doi.org/10.1038/nprot.2007.352

Nakamura E, Isobe H (2003) Functionalized fullerenes in water. The first 10 years of their chemistry, biology, and nanoscience. Acc Chem Res 36:807–815. https://doi.org/10.1021/ar030027y

Nakamura E, Isobe H, Tomita N et al (2000) Functionalized fullerene as an artificial vector for transfection. Angew Chem 39:4254–4257. https://doi.org/10.1002/1521-3773(20001201)39:23<4254::AID-ANIE4254>3.0.CO;2-O

Namgung R, Singha K, Yu MK et al (2010) Hybrid superparamagnetic iron oxide nanoparticle-branched polyethylenimine magnetoplexes for gene transfection of vascular endothelial cells. Biomaterials 31:4204–4213. https://doi.org/10.1016/j.biomaterials.2010.01.123

Namiki Y, Namiki T, Yoshida H et al (2009) A novel magnetic crystal–lipid nanostructure for magnetically guided in vivo gene delivery. Nat Nanotechnol 4:598–606. https://doi.org/10.1038/nnano.2009.202

Nance E, Timbie K, Miller GW et al (2014) Non-invasive delivery of stealth, brain-penetrating nanoparticles across the blood−brain barrier using MRI-guided focused ultrasound. J Control Release 189:123–132. https://doi.org/10.1016/j.jconrel.2014.06.031

Negishi Y, Omata D, Iijima H et al (2010) Enhanced laminin-derived peptide AG73-mediated liposomal gene transfer by bubble liposomes and ultrasound. Mol Pharm 7:217–226. https://doi.org/10.1021/mp900214s

Neumann E, Schaefer-Ridder M, Wang Y, Hofschneider PH (1982) Gene transfer into mouse lyoma cells by electroporation in high electric fields. EMBO J 1:841–845

Neves V, Heister E, Costa S et al (2012) Design of double-walled carbon nanotubes for biomedical applications. Nanotechnology 23:365102. https://doi.org/10.1088/0957-4484/23/36/365102

Nomikou N, McHale AP (2012) Microbubble-enhanced ultrasound-mediated gene transfer – Towards the development of targeted gene therapy for cancer. Int J Hyperthermia 28:300–310. https://doi.org/10.3109/02656736.2012.659235

Nomikou N, Tiwari P, Trehan T et al (2012) Studies on neutral, cationic and biotinylated cationic microbubbles in enhancing ultrasound-mediated gene delivery in vitro and in vivo. Acta Biomater 8:1273–1280. https://doi.org/10.1016/j.actbio.2011.09.010

Ogris M, Brunner S, Schüller S et al (1999) PEGylated DNA/transferrin-PEI complexes: reduced interaction with blood components, extended circulation in blood and potential for systemic gene delivery. Gene Ther 6:595–605. https://doi.org/10.1038/sj.gt.3300900

Ogura M, Sato S, Nakanishi K et al (2004) In vivo targeted gene transfer in skin by the use of laser-induced stress waves. Lasers Surg Med 34:242–248. https://doi.org/10.1002/lsm.20024

Okuda T, Sugiyama A, Niidome T, Aoyagi H (2004) Characters of dendritic poly(L-lysine) analogues with the terminal lysines replaced with arginines and histidines as gene carriers in vitro. Biomaterials 25:537–544

Omata D, Negishi Y, Hagiwara S et al (2011) Bubble liposomes and ultrasound promoted endosomal escape of TAT-PEG liposomes as gene delivery carriers. Mol Pharm 8:2416–2423. https://doi.org/10.1021/mp200353m

Omata D, Negishi Y, Hagiwara S et al (2012) Enhanced gene delivery using Bubble liposomes and ultrasound for folate-PEG liposomes. J Drug Target 20:355–363. https://doi.org/10.3109/1061186X.2012.660162

Ortner V, Kaspar C, Halter C et al (2012) Magnetic field-controlled gene expression in encapsulated cells. J Control Release 158:424–432. https://doi.org/10.1016/j.jconrel.2011.12.006

Owen J, Pankhurst Q, Stride E (2012) Magnetic targeting and ultrasound mediated drug delivery: benefits, limitations and combination. Int J Hyperthermia 28:362–373. https://doi.org/10.3109/02656736.2012.668639

Pantarotto D, Briand J-P, Prato M, Bianco A (2004) Translocation of bioactive peptides across cell membranes by carbon nanotubes. Chem Commun 0:16–17. doi: https://doi.org/10.1039/B311254C

Park JW, Bae KH, Kim C, Park TG (2011) Clustered magnetite nanocrystals cross-linked with PEI for efficient siRNA delivery. Biomacromolecules 12:457–465. https://doi.org/10.1021/bm101244j

Park HY, Noh EH, Chung H-M et al (2012) Efficient generation of virus-free iPS cells using liposomal magnetofection. PLoS One 7:e45812. https://doi.org/10.1371/journal.pone.0045812

Partha R, Lackey M, Hirsch A et al (2007) Self assembly of amphiphilic C60 fullerene derivatives into nanoscale supramolecular structures. J Nanobiotechnol 5:6. https://doi.org/10.1002/(sici)1521-3773(19990816)38:16<2403::aid-anie2403>3.0.co

Partha R, Mitchell LR, Lyon JL et al (2008) Buckysomes: fullerene-based nanocarriers for hydrophobic molecule delivery. ACS Nano 2:1950–1958. https://doi.org/10.1021/nn800422k

Pedrosa P, Vinhas R, Fernandes A, Baptista P (2015) Gold nanotheranostics: proof-of-concept or clinical tool? Nanomaterials 5:1853–1879. https://doi.org/10.3390/nano5041853

Pickard MR, Barraud P, Chari DM (2011) The transfection of multipotent neural precursor/stem cell transplant populations with magnetic nanoparticles. Biomaterials 32:2274–2284. https://doi.org/10.1016/j.biomaterials.2010.12.007

Plank C, Scherer F, Schillinger U, Anton M (2000) Magnetofection: enhancement and localization of gene delivery with magnetic particles under the influence of a magnetic field. J Gene Med 2 (Suppl):S24

Plank C, Zelphati O, Mykhaylyk O (2011) Magnetically enhanced nucleic acid delivery. Ten years of magnetofection—progress and prospects. Adv Drug Deliv Rev 63:1300–1331. https://doi.org/10.1016/j.addr.2011.08.002

Podesta JE, Al-Jamal KT, Herrero MA et al (2009) Antitumor activity and prolonged survival by carbon-nanotube-mediated therapeutic siRNA silencing in a human lung xenograft model. Small 5:1176–1185. https://doi.org/10.1002/smll.200801572

Prausnitz MR (2004) Microneedles for transdermal drug delivery. Adv Drug Deliv Rev 56:581–587. https://doi.org/10.1016/j.addr.2003.10.023

Praveen BB, Stevenson DJ, Antkowiak M et al (2011) Enhancement and optimization of plasmid expression in femtosecond optical transfection. J Biophotonics 4:229–235. https://doi.org/10.1002/jbio.201000105

Prijic S, Prosen L, Cemazar M et al (2012) Surface modified magnetic nanoparticles for immuno-gene therapy of murine mammary adenocarcinoma. Biomaterials 33:4379–4391. https://doi.org/10.1016/j.biomaterials.2012.02.061

Qiu Y, Luo Y, Zhang Y et al (2010) The correlation between acoustic cavitation and sonoporation involved in ultrasound-mediated DNA transfection with polyethylenimine (PEI) in vitro. J Control Release 145:40–48. https://doi.org/10.1016/j.jconrel.2010.04.010

Qiu Y, Zhang C, Tu J, Zhang D (2012) Microbubble-induced sonoporation involved in ultrasound-mediated DNA transfection in vitro at low acoustic pressures. J Biomech 45:1339–1345. https://doi.org/10.1016/j.jbiomech.2012.03.011

Raper SE, Chirmule N, Lee FS et al (2003) Fatal systemic inflammatory response syndrome in a ornithine transcarbamylase deficient patient following adenoviral gene transfer. Mol Genet Metab 80:148–158. https://doi.org/10.1016/j.ymgme.2003.08.016

Rejman J, Wagenaar A, Engberts JBFN, Hoekstra D (2004) Characterization and transfection properties of lipoplexes stabilized with novel exchangeable polyethylene glycol-lipid conjugates. Biochim Biophys Acta Biomembr 1660:41–52. https://doi.org/10.1016/j.bbamem.2003.10.011

Ren T, Song YK, Zhang G, Liu D (2000) Structural basis of DOTMA for its high intravenous transfection activity in mouse. Gene Ther 7:764–768. https://doi.org/10.1038/sj.gt.3301153

Rendall HA, Marchington RF, Praveen BB et al (2012) High-throughput optical injection of mammalian cells using a Bessel light beam. Lab Chip 12:4816–4820. https://doi.org/10.1039/c2lc40708f

Rezaee M, Oskuee RK, Nassirli H, Malaekeh-Nikouei B (2016) Progress in the development of lipopolyplexes as efficient non-viral gene delivery systems. J Control Release 236:1–14. https://doi.org/10.1016/j.jconrel.2016.06.023

Richardson CF, Schuster DI, Wilson SR (2000) Synthesis and characterization of water-soluble amino fullerene derivatives. Org Lett 2:1011–1014. https://doi.org/10.1021/ol990312a

Rojas-Chapana J, Troszczynska J, Firkowska I et al (2005) Multi-walled carbon nanotubes for plasmid delivery into *Escherichia coli* cells. Lab Chip 5:536–539. https://doi.org/10.1039/b500681c

Ryou S-M, Park M, Kim J-M et al (2011) Inhibition of xenograft tumor growth in mice by gold nanoparticle-assisted delivery of short hairpin RNAs against Mcl-1L. J Biotechnol 156:89–94. https://doi.org/10.1016/j.jbiotec.2011.07.037

Sanches PG, Grüll H, Steinbach OC (2011) See, reach, treat: ultrasound-triggered image-guided drug delivery. Ther Deliv 2:919–934

Sandre O, Genevois C, Garaio E et al (2017) In vivo imaging of local gene expression induced by magnetic hyperthermia. Genes (Basel) 8:61. https://doi.org/10.3390/genes8020061

Sanford JC, Klein TM, Wolf ED, Allen N (1987) Delivery of substances into cells and tissues using a particle bombardment process. Part Sci Technol 5:27–37. https://doi.org/10.1080/02726358708904533

Sapet C, Laurent N, de Chevigny A et al (2011) High transfection efficiency of neural stem cells with magnetofection. Biotechniques 50:187–189. https://doi.org/10.2144/000113628

Satoh Y, Kanda Y, Terakawa M et al (2005) Targeted DNA transfection into the mouse central nervous system using laser-induced stress waves. J Biomed Opt 10:60501. https://doi.org/10.1117/1.2128432

Sauer AM, de Bruin KG, Ruthardt N et al (2009) Dynamics of magnetic lipoplexes studied by single particle tracking in living cells. J Control Release 137:136–145. https://doi.org/10.1016/j.jconrel.2009.04.003

Scherer F, Plank C (2008) Magnetofection: using magnetic particles and magnetic force to enhance and to target nucleic acid delivery. In: Templeton NS (ed) Gene and cell therapy: therapeutic mechanisms and strategies, 3rd edn. CRC, Boca Raton, pp 379–404

Scherer F, Anton M, Schillinger U et al (2002) Magnetofection: enhancing and targeting gene delivery by magnetic force in vitro and in vivo. Gene Ther 9:102–109. https://doi.org/10.1038/sj.gt.3301624

Schwerdt A, Zintchenko A, Concia M et al (2008) Hyperthermia induced targeting of thermosensitive gene carriers to tumors. Hum Gene Ther 11:1283–1292. https://doi.org/10.1089/hgt.2008.064

Senior JH, Trimble KR, Maskiewicz R (1991) Interaction of positively-charged liposomes with blood: implications for their application in vivo. BBA Biomembr 1070:173–179. https://doi.org/10.1016/0005-2736(91)90160-A

Shan Y, Luo T, Peng C et al (2012) Gene delivery using dendrimer-entrapped gold nanoparticles as nonviral vectors. Biomaterials 33:3025–3035. https://doi.org/10.1016/j.biomaterials.2011.12.045

Shao P, Wang B, Wang Y et al (2011) The application of thermosensitive nanocarriers in controlled drug delivery. J Nanomater 2011:1–12. https://doi.org/10.1155/2011/389640

Sharma A, Tandon A, Tovey JCK et al (2011) Polyethylenimine-conjugated gold nanoparticles: gene transfer potential and low toxicity in the cornea. Nanomedicine 7:505–513. https://doi.org/10.1016/j.nano.2011.01.006

Shi Kam NW, Jessop TC, Wender PA, Dai H (2004) Nanotube molecular transporters: internalization of carbon nanotube–protein conjugates into mammalian cells. J Am Chem Soc 126:6850–6851. https://doi.org/10.1021/ja0486059

Shim MS, Kwon YJ (2012) Stimuli-responsive polymers and nanomaterials for gene delivery and imaging applications. Adv Drug Deliv Rev 64:1046–1059. https://doi.org/10.1016/j.addr.2012.01.018

Sigwalt D, Holler M, Iehl J et al (2011) Gene delivery with polycationic fullerene hexakis-adducts. Chem Commun 47:4640–4642. https://doi.org/10.1039/c0cc05783e

Sirsi SR, Hernandez SL, Zielinski L et al (2012) Polyplex-microbubble hybrids for ultrasound-guided plasmid DNA delivery to solid tumors. J Control Release 157:224–234. https://doi.org/10.1016/j.jconrel.2011.09.071

Sitharaman B, Zakharian TY, Saraf A et al (2008) Water-soluble fullerene (C60) derivatives as nonviral gene-delivery vectors. Mol Pharm 5:567–578. https://doi.org/10.1021/mp700106w
Soman P, Zhang W, Umeda A et al (2011) Femtosecond laser-assisted optoporation for drug and gene delivery into single mammalian cells. J Biomed Nanotechnol 7:334–341
Song HP, Yang JY, Lo SL et al (2010) Gene transfer using self-assembled ternary complexes of cationic magnetic nanoparticles, plasmid DNA and cell-penetrating Tat peptide. Biomaterials 31:769–778. https://doi.org/10.1016/j.biomaterials.2009.09.085
Stamatatos L, Leventis R, Zuckermann MJ, Silvius JR (1988) Interactions of cationic lipid vesicles with negatively charged phospholipid vesicles and biological membranes. Biochemistry 27:3917–3925. https://doi.org/10.1021/bi00411a005
Steitz B, Hofmann H, Kamau SW et al (2007) Characterization of PEI-coated superparamagnetic iron oxide nanoparticles for transfection: size distribution, colloidal properties and DNA interaction. J Magn Magn Mater 311:300–305. https://doi.org/10.1016/j.jmmm.2006.10.1194
Stern JM, Kibanov Solomonov VV, Sazykina E et al (2016) Initial evaluation of the safety of nanoshell-directed photothermal therapy in the treatment of prostate disease. Int J Toxicol 35:38–46. https://doi.org/10.1177/1091581815600170
Stevenson DJ, Gunn-Moore FJ, Campbell P, Dholakia K (2010) Single cell optical transfection. J R Soc Interface 7:863–871. https://doi.org/10.1098/rsif.2009.0463
Stride E, Porter C, Prieto AG, Pankhurst Q (2009) Enhancement of microbubble mediated gene delivery by simultaneous exposure to ultrasonic and magnetic fields. Ultrasound Med Biol 35:861–868. https://doi.org/10.1016/j.ultrasmedbio.2008.11.010
Sugano M, Negishi Y, Endo-Takahashi Y et al (2012) Gene delivery system involving Bubble liposomes and ultrasound for the efficient in vivo delivery of genes into mouse tongue tissue. Int J Pharm 422:332–337. https://doi.org/10.1016/j.ijpharm.2011.11.001
Suk JS, Xu Q, Kim N et al (2016) PEGylation as a strategy for improving nanoparticle-based drug and gene delivery. Adv Drug Deliv Rev 99:28–51. https://doi.org/10.1016/j.addr.2015.09.012
Sun S-L, Lo Y-L, Chen H-Y, Wang L-F (2012) Hybrid polyethylenimine and polyacrylic acid-bound iron oxide as a magnetoplex for gene delivery. Langmuir 28:3542–3552. https://doi.org/10.1021/la204529u
Suzuki R, Namai E, Oda Y et al (2010) Cancer gene therapy by IL-12 gene delivery using liposomal bubbles and tumoral ultrasound exposure. J Control Release 142:245–250. https://doi.org/10.1016/j.jconrel.2009.10.027
Suzuki R, Oda Y, Utoguchi N, Maruyama K (2011) Progress in the development of ultrasound-mediated gene delivery systems utilizing nano- and microbubbles. J Control Release 149:36–41. https://doi.org/10.1016/j.jconrel.2010.05.009
Tang Q, Zhang D, Cong X et al (2008) Using thermal energy produced by irradiation of Mn-Zn ferrite magnetic nanoparticles (MZF-NPs) for heat-inducible gene expression. Biomaterials 29:2673–2679. https://doi.org/10.1016/j.biomaterials.2008.01.038
Tang LAL, Wang J, Loh KP (2010a) Graphene-based SELDI probe with ultrahigh extraction and sensitivity for DNA oligomer. J Am Chem Soc 132:10976–10977. https://doi.org/10.1021/ja104017y
Tang Z, Wu H, Cort JR et al (2010b) Constraint of DNA on functionalized graphene improves its biostability and specificity. Small 6:1205–1209. https://doi.org/10.1002/smll.201000024
Tencomnao T, Apijaraskul A, Rakkhithawatthana V et al (2011) Gold/cationic polymer nano-scaffolds mediated transfection for non-viral gene delivery system. Carbohydr Polym 84:216–222. https://doi.org/10.1016/j.carbpol.2010.12.063
Terakawa M, Ogura M, Sato S et al (2004) Gene transfer into mammalian cells by use of a nanosecond pulsed laser-induced stress wave. Opt Lett 29:1227–1229
Thierry AR, Rabinovich P, Peng B et al (1997) Characterization of liposome-mediated gene delivery: expression, stability and pharmacokinetics of plasmid DNA. Gene Ther 4:226–237. https://doi.org/10.1038/sj.gt.3300350
Timbie KF, Mead BP, Price RJ (2015) Drug and gene delivery across the blood–brain barrier with focused ultrasound. J Control Release 219:61–75. https://doi.org/10.1016/j.jconrel.2015.08.059

Tirlapur UK, König K (2002) Cell biology: targeted transfection by femtosecond laser. Nature 418:290–291. https://doi.org/10.1038/418290a

Tokuyama H, Yamago S, Nakamura E et al (1993) Photoinduced biochemical activity of fullerene carboxylic acid. J Am Chem Soc 115:7918–7919. https://doi.org/10.1021/ja00070a064

Tsampoula X, Garcés-Chávez V, Comrie M et al (2007) Femtosecond cellular transfection using a nondiffracting light beam. Appl Phys Lett 91:53902. https://doi.org/10.1063/1.2766835

Tseng WC, Haselton FR, Giorgio TD (1997) Transfection by cationic liposomes using simultaneous single cell measurements of plasmid delivery and transgene expression. J Biol Chem 272:25641–25647

Uchugonova A, König K, Bueckle R et al (2008) Targeted transfection of stem cells with sub-20 femtosecond laser pulses. Opt Express 16:9357–9364

Un K, Kawakami S, Suzuki R et al (2010) Development of an ultrasound-responsive and mannose-modified gene carrier for DNA vaccine therapy. Biomaterials 31:7813–7826. https://doi.org/10.1016/j.biomaterials.2010.06.058

Un K, Kawakami S, Yoshida M et al (2011a) The elucidation of gene transferring mechanism by ultrasound-responsive unmodified and mannose-modified lipoplexes. Biomaterials 32:4659–4669. https://doi.org/10.1016/j.biomaterials.2011.03.013

Un K, Kawakami S, Suzuki R et al (2011b) Suppression of melanoma growth and metastasis by DNA vaccination using an ultrasound-responsive and mannose-modified gene carrier. Mol Pharm 8:543–554. https://doi.org/10.1021/mp100369n

Vainauska D, Kozireva S, Karpovs A et al (2012) A novel approach for nucleic acid delivery into cancer cells. Medicina (B Aires) 48:324–329

van der Aa MAEM, Mastrobattista E, Oosting RS et al (2006) The nuclear pore complex: the gateway to successful nonviral gene delivery. Pharm Res 23:447–459. https://doi.org/10.1007/s11095-005-9445-4

van der Maaden K, Jiskoot W, Bouwstra J (2012) Microneedle technologies for (trans)dermal drug and vaccine delivery. J Control Release 161:645–655. https://doi.org/10.1016/j.jconrel.2012.01.042

Vandenbroucke RE, Lentacker I, Demeester J et al (2008) Ultrasound assisted siRNA delivery using PEG-siPlex loaded microbubbles. J Control Release 126:265–273. https://doi.org/10.1016/j.jconrel.2007.12.001

Varkouhi AK, Foillard S, Lammers T et al (2011) SiRNA delivery with functionalized carbon nanotubes. Int J Pharm 416:419–425. https://doi.org/10.1016/j.ijpharm.2011.02.009

Vlaskou D, Mykhaylyk O, Krötz F et al (2010a) Magnetic and acoustically active lipospheres for magnetically targeted nucleic acid delivery. Adv Funct Mater 20:3881–3894. https://doi.org/10.1002/adfm.200902388

Vlaskou D, Pradhan P, Bergemann C et al (2010b) Magnetic microbubbles: magnetically targeted and ultrasound-triggered vectors for gene delivery in vitro. In: AIP conference proceedings. pp 485–494

Vlaskou D, Plank C, Mykhaylyk O (2013) Magnetic and acoustically active microbubbles loaded with nucleic acids for gene delivery. In: Nanotechnology for nucleic acid delivery. Humana, Totowa, NJ, pp 205–241

Wahajuddin A (2012) Superparamagnetic iron oxide nanoparticles: magnetic nanoplatforms as drug carriers. Int J Nanomed 7:3445–3471. https://doi.org/10.2147/IJN.S30320

Walther W, Stein U (2009) Heat-responsive gene expression for gene therapy. Adv Drug Deliv Rev 61:641–649. https://doi.org/10.1016/j.addr.2009.02.009

Wang Y, Li Z, Hu D et al (2010) Aptamer/graphene oxide nanocomplex for in situ molecular probing in living cells. J Am Chem Soc 132:9274–9276. https://doi.org/10.1021/ja103169v

Wang C, Ding C, Kong M et al (2011) Tumor-targeting magnetic lipoplex delivery of short hairpin RNA suppresses IGF-1R overexpression of lung adenocarcinoma A549 cells in vitro and in vivo. Biochem Biophys Res Commun 410:537–542. https://doi.org/10.1016/j.bbrc.2011.06.019

Wang Y, Li Z, Weber TJ et al (2013) In situ live cell sensing of multiple nucleotides exploiting DNA/RNA aptamers and graphene oxide nanosheets. Anal Chem 85:6775–6782. https://doi.org/10.1021/ac400858g

Wells DJ (2004) Gene therapy progress and prospects: electroporation and other physical methods. Gene Ther 11:1363–1369. https://doi.org/10.1038/sj.gt.3302337

Wen Y, Xing F, He S et al (2010) A graphene-based fluorescent nanoprobe for silver(I) ions detection by using graphene oxide and a silver-specific oligonucleotide. Chem Commun 46:2596–2598. https://doi.org/10.1039/b924832c

Wijaya A, Schaffer SB, Pallares IG, Hamad-Schifferli K (2009) Selective release of multiple DNA oligonucleotides from gold nanorods. ACS Nano 3:80–86. https://doi.org/10.1021/nn800702n

Winkler J (2015) Therapeutic oligonucleotides with polyethylene glycol modifications. Future Med Chem 7:1721–1731. https://doi.org/10.4155/fmc.15.94

Wolff JA, Malone RW, Williams P et al (1990) Direct gene transfer into mouse muscle in vivo. Science 247:1465–1468

Wu GY, Wu CH (1987) Receptor-mediated in vitro gene transformation by a soluble DNA carrier system. J Biol Chem 262:4429–4432

Wu M, Kempaiah R, Huang P-JJ et al (2011) Adsorption and desorption of DNA on graphene oxide studied by fluorescently labeled oligonucleotides. Langmuir 27:2731–2738. https://doi.org/10.1021/la1037926

Xie L, Jiang Q, He Y et al (2015) Insight into the efficient transfection activity of a designed low aggregated magnetic polyethyleneimine/DNA complex in serum-containing medium and the application in vivo. Biomater Sci 3:446–456. https://doi.org/10.1039/C4BM00317A

Xu J, Yudasaka M, Kouraba S et al (2008) Single wall carbon nanohorn as a drug carrier for controlled release. Chem Phys Lett 461:189–192. https://doi.org/10.1016/j.cplett.2008.06.077

Yamagata M, Kawano T, Shiba K et al (2007) Structural advantage of dendritic poly(l-lysine) for gene delivery into cells. Bioorg Med Chem 15:526–532. https://doi.org/10.1016/j.bmc.2006.09.033

Yamaguchi M, Ito A, Okamoto N et al (2012) Heat-inducible transgene expression system incorporating a positive feedback loop of transcriptional amplification for hyperthermia-induced gene therapy. J Biosci Bioeng 114:460–465. https://doi.org/10.1016/j.jbiosc.2012.05.006

Yang S-Y, Sun J-S, Liu C-H et al (2008) Ex vivo magnetofection with magnetic nanoparticles: a novel platform for nonviral tissue engineering. Artif Organs 32:195–204. https://doi.org/10.1111/j.1525-1594.2007.00526.x

Yao C-P, Zhang Z-X, Rahmanzadeh R, Huettmann G (2008) Laser-based gene transfection and gene therapy. IEEE Trans Nanobiosci 7:111–119. https://doi.org/10.1109/TNB.2008.2000742

Yoo D, Lee J-H, Shin T-H, Cheon J (2011) Theranostic magnetic nanoparticles. Acc Chem Res 44:863–874. https://doi.org/10.1021/ar200085c

Yoon CS, Park JH (2010) Ultrasound-mediated gene delivery. Expert Opin Drug Deliv 7:321–330. https://doi.org/10.1517/17425241003596329

Young KL, Scott AW, Hao L et al (2012) Hollow spherical nucleic acids for intracellular gene regulation based upon biocompatible silica shells. Nano Lett 12:3867–3871. https://doi.org/10.1021/nl3020846

Yudina A, Moonen C (2012) Ultrasound-induced cell permeabilisation and hyperthermia: strategies for local delivery of compounds with intracellular mode of action. Int J Hyperthermia 28:311–319. https://doi.org/10.3109/02656736.2012.664307

Yudina A, de Smet M, Lepetit-Coiffé M et al (2011) Ultrasound-mediated intracellular drug delivery using microbubbles and temperature-sensitive liposomes. J Control Release 155:442–448. https://doi.org/10.1016/j.jconrel.2011.06.006

Yudina A, Lepetit-Coiffé M, De Smet M et al (2012) In vivo temperature controlled ultrasound-mediated intracellular delivery of cell-impermeable compounds. J Control Release 161:90–97. https://doi.org/10.1016/j.jconrel.2012.04.018

Zabner J, Fasbender AJ, Moninger T et al (1995) Cellular and molecular barriers to gene transfer by a cationic lipid. J Biol Chem 270:18997–19007. https://doi.org/10.1074/jbc.270.32.18997

Zeira E, Manevitch A, Manevitch Z et al (2007) Femtosecond laser: a new intradermal DNA delivery method for efficient, long-term gene expression and genetic immunization. FASEB J 21:3522–3533. https://doi.org/10.1096/fj.06-7528com

Zelphati O, Uyechi LS, Barron LG, Szoka FC (1998) Effect of serum components on the physico-chemical properties of cationic lipid/oligonucleotide complexes and on their interactions with cells. Biochim Biophys Acta Lipids Lipid Metab 1390:119–133. https://doi.org/10.1016/S0005-2760(97)00169-0

Zheng D, Giljohann DA, Chen DL et al (2012) Topical delivery of siRNA-based spherical nucleic acid nanoparticle conjugates for gene regulation. Proc Natl Acad Sci U S A 109:11975–11980. https://doi.org/10.1073/pnas.1118425109

Zhou X, Huang L (1994) DNA transfection mediated by cationic liposomes containing lipopolylysine: characterization and mechanism of action. Biomembranes 1189:195–203

Zhou X-F, Liu B, Yu X-H et al (2007) Using magnetic force to enhance immune response to DNA vaccine. Small 3:1707–1713. https://doi.org/10.1002/smll.200700151

Zhou X, Laroche F, Lamers GEM et al (2012) Ultra-small graphene oxide functionalized with polyethylenimine (PEI) for very efficient gene delivery in cell and zebrafish embryos. Nano Res 5:703–709. https://doi.org/10.1007/s12274-012-0254-x

Chapter 3
Specific Systems for Imaging

Chi Hong Sum, Samantha Marisha Shortall, Jessica Antoinetta Nicastro, and Roderick Slavcev

Contents

Abstract Microscopy allows for the characterization of small objects invisible to the naked eye, a technique that, since its conception, has played a key role in the development across nearly every field of science and technology. Given the nanometer size of the materials explored in the field of nanotechnology, the contributions of modern microscopes that can visualize these materials are indispensable, and the ever-improving technology is paramount to the future success of the field. This chapter will focus on four fundamental areas of microscopy used in the field of nanotechnology including fluorescence microscopy (Sect. 3.1), particle tracking and photoactivated localization microscopy (Sect. 3.2), quantum dots and fluorescence resonance energy transfer (Sect. 3.3), and cellular MRI and PET labeling (Sect. 3.4). The functionality, as well as the current and recommended usage of each given imaging system, will be discussed.

Keywords Fluorescence microscopy · Multiphoton microscopy · Total internal reflection fluorescence microscopy · Single particle tracking · Quantum dot · Fluorescence resonance energy transfer · Cellular MRI · PET labeling

C. H. Sum · S. M. Shortall · J. A. Nicastro · R. Slavcev (✉)
University of Waterloo, Waterloo, ON, Canada
e-mail: slavcev@uwaterloo.ca

© Springer International Publishing AG, part of Springer Nature 2018
R. A. Slavcev et al. (eds.), *Nanomedicine*, Experientia Supplementum 110,
https://doi.org/10.1007/978-3-319-78259-1_3

3.1 Fluorescence Microscopy

Fluorescence microscopy is a rapidly expanding technique, which has made the identification of specific cellular components and functions possible. This form of microscopy depends on the ability of a given specimen to emit energy in such a way that it can be detected as visible light when irradiated with light from the given instrument at a specific wavelength. The fluorescence microscope apparatus sends out an excitation light specific to the desired wavelength in the specimen and sorts out the differentiation in light emissions to detect the weaker light and form an image. The areas with fluorescing light can be observed in the microscope and will be contrasted against a dark background. The areas of fluorescence microscopy to be reviewed will include multiphoton microscopy and total internal reflection fluorescence microscopy (TIRFM).

3.1.1 Multiphoton Microscopy

Multiphoton microscopy is a leading technology in the exploration of biological machinery in live organisms, particularly for the visualization of real-time cellular structure and function. This technology provides 3D optical sectioning, similar to confocal microscopy, but with significantly less damage to the host tissue. Multiphoton microscopy has been exploited for the imaging of living specimens and is an ideal tool for the real-time analysis of cellular function in neural networks (Svoboda and Yasuda 2006), microvasculature (Chen et al. 2009a), and developmental studies (Hoover and Squier 2013; Chu et al. 2003).

Traditional confocal fluorescence microscopy is based on a linear effect, where one photon from a specimen is absorbed to constitute the emission of a single fluorescent photon by the same fluorescent molecule (Fig. 3.1). The resultant fluorescence intensity is in direct correlation with the excitation power (Christensen and Nedergaard 2011; Ustione and Piston 2011). In contrast, multiphoton fluorescence microscopy employs the simultaneous absorption of two or more photons to provide sufficient energy required for excitation (Fig. 3.1). In the case of two-photon-excited fluorescence microscopy (TPEF), excitation as a result of simultaneous absorption of two photons, each possessing approximately half the required energy and twice the wavelength, corresponds to fluorescence emission identical to emissions derived from single-photon excitation. Such nonlinear interactions in multiphoton microscopy are rare events as the probability of two photons interacting simultaneously with a single fluorescent molecule is very low. The application of continuous wave lasers for multiphoton microscopy is not warranted as they generate steady streams of photons that are evenly distributed across space and time (Fig. 3.2). Instead, pulsed femtosecond lasers are used as they yield high photon density by forcing a high number of photons to be simultaneously delivered within a confined region of interest (Fig. 3.2) (Christensen and Nedergaard 2011; Mojzisova and Vermot 2011). As such, fluorescence excitation is localized at a specific focal point where the photons are concentrated. The employment of photons possessing higher wavelengths, at infrared

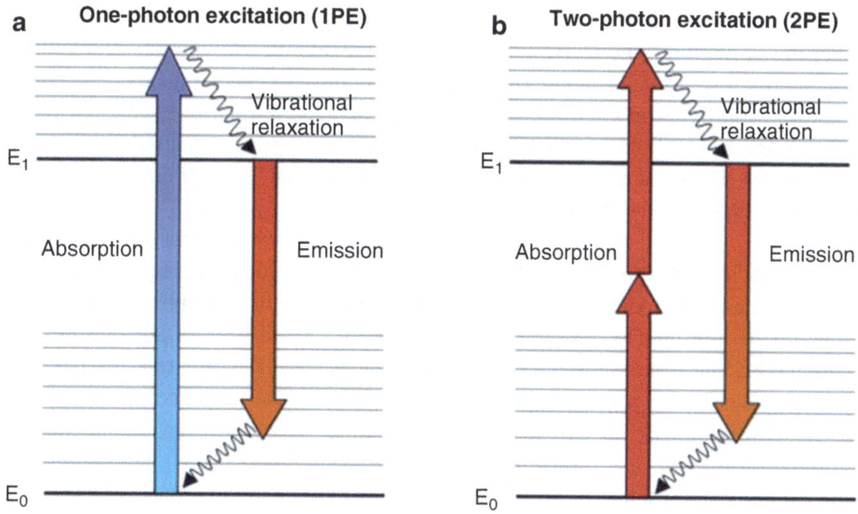

Fig. 3.1 Simplified Jablonski diagrams highlighting the differences between the excitation of one photon and two photons. Adapted from Paoli et al. (2009)

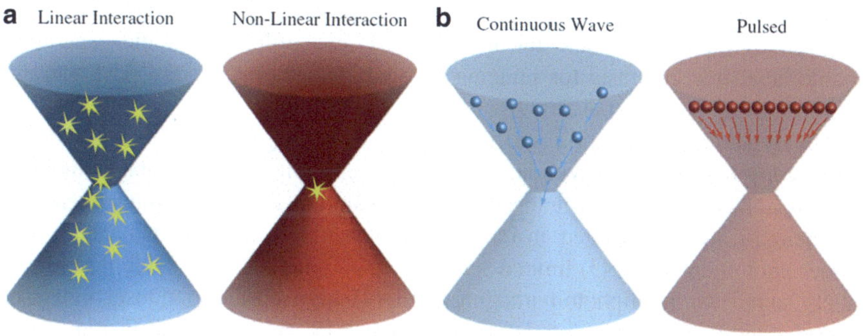

Fig. 3.2 (**a**) One-Photon excitation (1PE); and; (**b**) Two-photon excitation (2PE). Adapted from Christensen and Nedergaard (2011)

(IR) or near-infrared (NIR) ranges, reduces scattering since photon scattering is inversely proportional to wavelength of the fourth power (λ^4) (Mojzisova and Vermot 2011). Penetration depth is enhanced as NIR and IR light lies within the optical window of biological tissues at which intrinsic absorption of living cells and tissues is several orders of magnitude lowered than that of visible light. Penetration peaks at wavelengths around 1200 nm, just prior to significant absorption by water at 1300 nm (Mojzisova and Vermot 2011).

Multiphoton microscopy possesses several inherent advantages over traditional confocal microscopy. Superior penetration achieved by NIR and IR light increases the maximum imaging depth from tens of microns up to a recorded 1.6 mm (Kobat et al. 2011; Andresen et al. 2009). Two-photon microscopy has been shown to be able to image sixfold deeper than confocal microscopy when imaging the same sample using the same fluorophores (Ustione and Piston 2011). In conjunction with submicron spatial resolutions, capable of imaging subcellular components and fine tissue structures, multiphoton microscopy is able to observe cell migration, cell-cell interactions, and intracellular signaling within dense tissues of live animals (Andresen et al. 2009). Reduced photon absorption of NIR and IR light dramatically limits phototoxicity from the generation of reactive oxygen species. This feature has enabled the imaging technology to visualize living tissues, including developmental studies in living embryos. Since fluorescence excitation is localized at a specific focal point, collateral photodamage from out-of-focus fluorescence and photobleaching is limited (Christensen and Nedergaard 2011; Mojzisova and Vermot 2011; Andresen et al. 2009; Hayek et al. 2007; Tkaczyk and Tkaczyk 2011; Wang et al. 2010); this corresponds to highly sensitive detection of multiphoton fluorescence and the potential for observations over extended periods of time. The nonlinear dependence of multiphoton microscopy allows for fine control of the lateral (x-y plane) and axial dimensions (z-axis), a feature absent in traditional confocal fluorescence microscopy. The intrinsic properties of optical sectioning and 3D spatial localization permit 3D imaging from reconstruction of overlaid multiphoton fluorescence (Mojzisova and Vermot 2011; Hayek et al. 2007; Li et al. 2009). As multiphoton lasers are tunable to emit specific wavelengths, a variety of fluorophores can be excited using a single laser (Fritzky and Lagunoff 2013). Lastly, the excitation of UV absorption bands of endogenous fluorophores and the induction of second/third harmonic generation enable label-free contrasts for imaging by multiphoton microscopy (Ustione and Piston 2011; Andresen et al. 2009).

The dynamic investigation into the field of living organisms with multiphoton microscopy has not come without some drawbacks. These disadvantages include, but are not limited to, (1) reduced spatial resolution from increasing excitation wavelengths, (2) accelerated in-focus photobleaching and photodamage due to higher photon densities, (3) limited selection of commercially available fluorescent labels suited for multiphoton imaging, and (4) high cost of femtosecond laser systems and appropriate optics corrected for NIR (Ustione and Piston 2011; Andresen et al. 2009; Hayek et al. 2007; Tkaczyk and Tkaczyk 2011; Christensen et al. 2001). Despite the improved imaging depth with multiphoton microscopy, it is still unable to generate sufficient penetration depth in optically dense tissues; this severely hampers the applicability of multiphoton microscopy for in vivo, deep tissue/organ imaging.

With respect to gene delivery, multiphoton microscopy has been applied to many different aspects throughout the delivery process. The functionalization of two-photon absorbing fluorophores to polyethylenimine (PEI), through conjugation to amine groups of PEI, has made it possible for monitoring intracellular trafficking of PEI/DNA polyplexes by two-photon microscopy. Using the conjugated PEI, Hayek et al. (2007) were able to demonstrate transfection efficiencies comparable to commercially available JetPEI/DNA polyplexes at 10:1 N^{+}/P^{-} charge ratio; the accumulation of theses polyplexes around the perinuclear region suggests that the conjugation of fluorophores did not compromise internalization nor the intracellular trafficking of the polyplexes. However, as the conjugation takes place at the amine

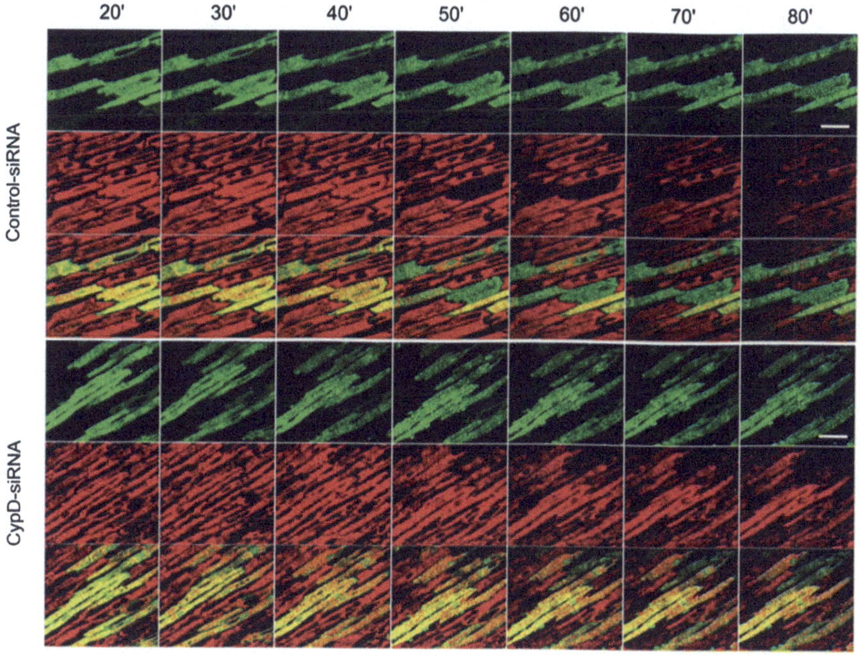

Fig. 3.3 Representative images of real-time monitoring of membrane potential in perfused rat hearts during 20–80 min after ischemia. For each group (control-siRNA and CydD-siRNA), the three panels represent GFP fluorescence (upper panel), TMRE fluorescence (middle panel), and merged images (lower panel). Adapted from (Kato et al. 2009)

groups of PEI, slight interferences with DNA compaction are evident by high fluorescence levels from YOYO-1 labeled DNA after complexation along with larger polyplex sizes at lower charge ratios.

The benefits of increased penetration depth using multiphoton microscopy are highlighted by the in vivo imaging of live animals and animal models. Based on evidence from two-photon imaging, Kato et al. (2009) demonstrate the potential of RNA interference of cyclophilin D (CydD) for cardioprotective therapy (Fig. 3.3). The adenovirus-mediated co-delivery of CydD-siRNA and GFP genes into cardiomyocytes (via intramyocardial injection into rat heart) and subsequent administration of tetramethylrhodamine ethyl ester (TMRE) to excised Langendorff-perfused rat heart permit the assessment of CydD-siRNA delivery along with its effects on membrane potential by two-photon microscopy. Two-photon microscopy is able to effectively monitor dynamic spatiotemporal changes in membrane potential of living heart models under ischemic condition. Gene silencing of CydD, a putative component of mitochondria permeability transition pore (MPTP), has been shown to prevent the progressive loss of mitochondrial membrane potential associated with cardiomyocyte cell death upon myocardial ischemia. Such effect is confirmed by the steady TMRE fluorescence achieved in GFP-positive cells, in conjunction with significantly reduced TMRE fluorescence in GFP negative cells,

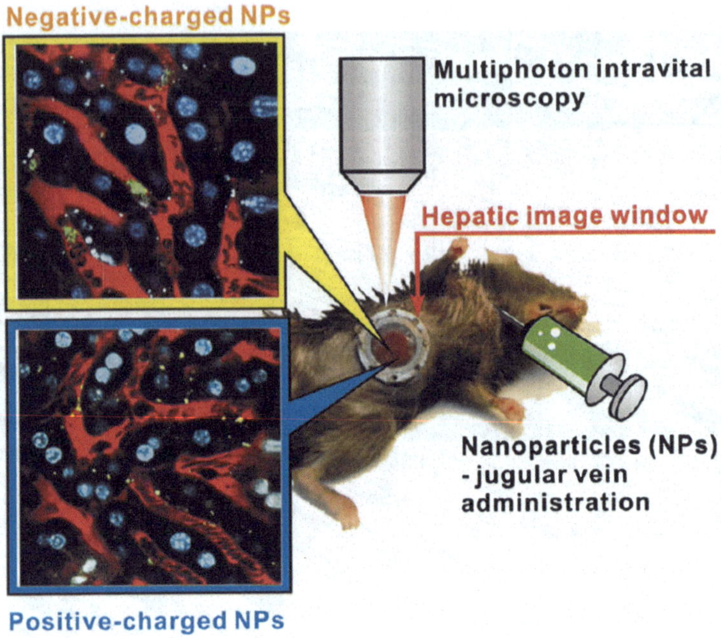

Fig. 3.4 Intravital multiphoton imaging through a hepatic imaging chamber. Adapted from Cheng et al. (2012)

from real-time two-photon imaging of perfused rat heart (see Fig. 3.4) (Kato et al. 2009). The study of cellular events in vivo is better achieved by intravital multiphoton imaging due to the high spatial and temporal resolutions not seen with whole-body imaging. Through intravital multiphoton imaging, Cheng et al. (2012) evaluated the dynamics of fluorescently labeled mesoporous silica nanoparticle transport through a hepatic imaging chamber. Such application of multiphoton imaging proves to be advantageous as longer excitation wavelength permits extended time-lapse imaging with reduced photodamage; in addition, the simultaneous acquisition of fluorescence and second harmonic generation signals offer better insight on the transport of nanoparticles across various underlying biological structures. The authors also highlight the potential for using such imaging techniques in preclinical assessment of nanoparticles, like gene delivery vectors, with various surface modifications (Cheng et al. 2012).

The application of gene delivery and subsequent imaging of gene products has been imaged by multiphoton imaging to study intracellular protein dynamics and cellular processes. Martini et al. (2007) demonstrated a novel approach to quantitatively monitor dynamic processes in vivo through selective two-photon fluorescence activation accompanied by real-time two-photon microscopy. Using this approach, the authors demonstrated quantitative monitoring of intracellular shuttling of *Arabidopsis* MYB transcription factor LHY/CCA1-like 1 (LCL1), a transcription factor possessing both nuclear localization and nuclear export signals, from nucleus to cytoplasm. This is achieved through the transfection of plasmids encoding photo-activatable GFP-LCL1 fusion protein. Transient expression of this fusion protein is imaged through selective, in-focus two-photon activation of photo-activatable GFP followed by its fluorescence emission upon excitation by either 488 nm light with one-photon laser scanning microscopy or 920 nm light with two-photon laser scanning microscopy (Martini et al. 2007). Through two-photon activation, selective labeling of the nucleus allows for quantitative assessment of the redistribution of photoactivated GFP into the cytoplasm.

As the endocytic pathway is one of the major pathways associated with vector internalization, Lysotrackers are often applied in co-localization experiments to track intracellular trafficking and subsequent escape of vectors from endosomal and endolysosomal compartments. Multiphoton microscopy is generally not considered for such experiments due to a lack of lysosomal two-photon probes with distinct emission bands differing from other two-photon probes. The development of lysosomal two-photon probes, CLT blue and CLT yellow (Son et al. 2011), permits in vivo co-localization experiments with noted visualization up to the depths of 80–170 μm in live tissues. The excitation with 750–840 nm light results in fluorescence emission, at 400–475 nm and 550–625 nm, by CLT blue and CLT yellow, respectively. These two-photon probes possess several desirable properties required by in vivo, live cell co-localization experiments including (1) efficient loading into cells and tissues, (2) negligible cytotoxicity, (3) selective localization to lysosomes, (4) high photostability, and (5) potential for extended time-lapse imaging (see Fig. 3.5) with minimal interference by pH (Son et al. 2011).

The utilization of multiphoton microscopy in vivo imaging has extended across many biological practices. One of the more recent studies, by Turcotte et al. (2016), applied the technology for a multi-scale assessment of the myelin structure. The group studied and developed models for myelination patterns as well as the screening of pro-myelination drugs in zebrafish spinal cords, an enquiry that was accomplished label-free with the combination of multiphoton imaging and polarization-resolved

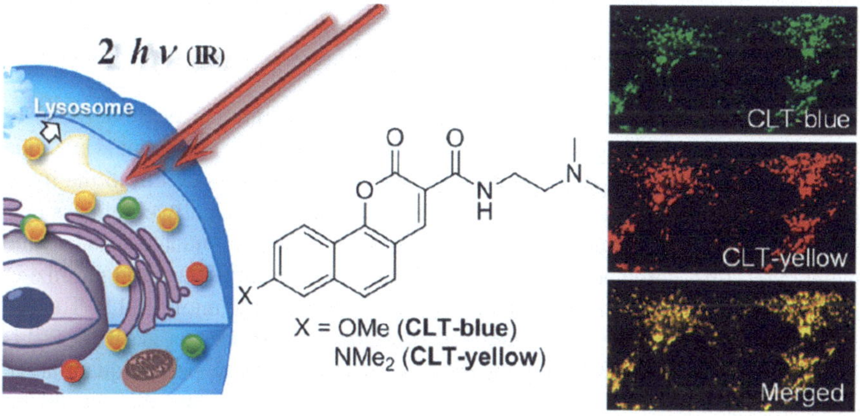

Fig. 3.5 Lysosomal two-photon probes, CLT blue and CLT yellow. Adapted from Son et al. (2011)

measurements (Turcotte et al. 2016). This combined technology, also called coherent anti-Stokes Raman scattering (CARS), functions by controlling the polarization of two different laser beams at two different wavelengths. This allows for label-free imaging that searches instead for single molecular vibrations of the intracellular molecules that match the energy difference of the pair of incoming photon laser beams (Turcotte et al. 2016; Bélanger et al. 2015). CARS imaging was a resource for visualizing the myelin structure in a way that provided novel insight into the myelin structure development and has been further applied to the insightful imaging of other biological systems (Turcotte et al. 2016; Bélanger et al. 2015; de Vito et al. 2012).

In conclusion, multiphoton microscopy offers numerous imaging benefits compared to single-photon confocal microscopy and has proven to be an ideal tool for in vivo imaging with its limited effect on cell viability. The technology has been applied to the imaging of gene delivery and gene products, intracellular components, and has been combined with polarization-resolved measurements for the label-free imaging of myelin sheaths among other biological products.

3.1.2 Total Internal Reflection Fluorescence Microscopy

Total internal reflection fluorescence microscopy (TIRFM), also known as evanescent field microscopy, relies on an illumination provided by an exponentially decaying fluorescence excitation field that is well suited to visualize fluorescently labeled products (Trexler and Taraska 2017; Steyer and Almers 1999). The resulting image is based on the total internal reflection (TIR) achieved when light propagates from a dense medium of high refractive index (n_2) into a less dense medium of lower refractive index (n_1) at incidence angles greater than the critical angle (Fritzky and Lagunoff 2013; Toomre and Manstein 2001). The critical angle is determined by the relative refractive indexes, whereby a small n_2/n_1 ratio constitutes a shallow critical angle for TIR; at angles greater than the critical angle, a portion of the incident light penetrates the less dense medium as an evanescent wave that propagates parallel to the interface at short distances, ranging from 100 to 200 nm, into adherent cells (Fig. 3.6) (Toomre and Manstein 2001; Parhamifar and Moghimi 2012; Mattheyses et al. 2010).

TIRFM has been extensively used in observing membrane processes due to several inherent advantages. The limited penetration of the evanescent wave, due to exponential decreases in light intensity with increasing distances from the interface, confines imaging at the plasma membrane which significantly eliminates out-of-focus light and excludes interferences from fluorophores situated distal to the membrane; this corresponds to better signal-to-noise ratios while minimizing photobleaching and cellular photodamage (Fritzky and Lagunoff 2013; Toomre and Manstein 2001; Parhamifar and Moghimi 2012; Mattheyses et al. 2010, 2011; Mattheyses and Axelrod 2006). Such decreases in intensity permit quantitative calculation of particle movement upon fluorophore excitation by the evanescent wave. TIRFM is a complementary technique that is compatible with fluorescence recovery after photobleaching (FRAP) and Förster resonance energy transfer

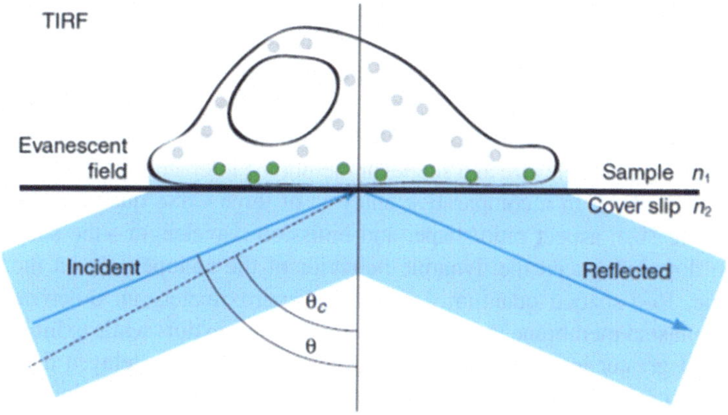

Fig. 3.6 The principle of TIR fluorescence. Adapted from Mattheyses et al. (2010)

(FRET) while being applicable with other microscopy techniques including epifluorescence microscopy, confocal microscopy, atomic force microscopy (AFM), fluorescence lifetime imaging microscopy (FLIM), and fluorescence correlation spectroscopy (FCS) (Brandenburg and Zhuang 2007). In particular, the combination of TIRFM and FCS offers quantitative data pertaining to surface binding, surface diffusion coefficients, and average number of fluorophores per surface area (Fritzky and Lagunoff 2013).

TIRFM has been implemented for the assessment of protein trafficking, membrane dynamics, endocytosis, exocytosis, and virus movement (Steyer and Almers 1999; Mattheyses et al. 2011; Brandenburg and Zhuang 2007). Recently, Trexler and Taraska used TIRFM to demonstrate a comprehensive method to image exocytosis in living endocrine cells, including the local protein dynamics at the site (Trexler and Taraska 2017). The group used TIRF with two colors to identify the individual and average behavior of molecules at the sites of exocytosis, a method that could help to describe the role of these proteins in exocytic fusion in mammalian cells (Trexler and Taraska 2017). Jouvenet et al. showed another example of the use of TIRFM in cellular mechanisms, when the group imaged the biogenesis of HIV-1 virus-like particles (VLPs) in live HeLa cells (Jouvenet et al. 2008). The group used fluorescently labeled VLPs imaged with TIRFM to visualize the virus assembly process including the role of one of the viral structural proteins, Gag. The technology provided a method for imaging a previously unobserved view of the viral genesis. The authors, however, noted a disadvantage of the imaging platform where any possible VLP assembly in the internal membranes of the cell would be out of the view afforded by TIRFM (Jouvenet et al. 2008). The insight gained in biological systems from the use of TIRFM is irreplaceable, and the imaging platform has a great deal of potential in the imaging of new technologies, including gene therapy.

With respect to gene delivery, TIRFM enables the analysis of nanoparticle internalization as well as ligand-receptor interactions upon targeted delivery by receptor-mediated endocytosis. Using TIRFM and fluorescent fusion proteins, clathrin-ds-Red and dynamin-1/2-GFP, the dynamics of clathrin and dynamin are analyzed with respect to the physical events of endocytosis including the scission of vesicle from plasma membrane and the early stages of vesicle internalization (Mattheyses et al. 2011). This validates the application of TIRFM for the evaluation of nanoparticle internalization during gene delivery. Through TIRFM and hyperspectral confocal fluorescence (HCF), Aaron et al. indicated the potential influence of nanoparticle size and shape on membrane interaction and internalization (Aaron et al. 2011). In a comparative analysis of three CdSe quantum dots—each with varying size, aspect ratio/shape, and emission wavelength—the shape of the quantum dots influenced the dynamic behavior of the nanoparticle on the plasma membrane. Rod-shaped quantum dots diffuse approximately an order magnitude slower in plasma membrane than most spherical quantum dots while exhibiting two- to threefold greater internalization duration as determined by delayed decreases in signal intensities. With respect to size, results indicate a positive correlation between increasing nanoparticle size and increasing rates of endocytosis/lysosomal sorting (Aaron et al. 2011). For quantitative characterization of the efficacy and kinetics of internalization, the application of flow cytometry with TIRF (TIRF-FC) permits rapid assessment of nanoparticle uptake on large cell populations. Although TIRF-FC lacked the spatial resolution to observe membrane processes associated with internalization, Wang et al. have confirmed that decreases in fluorescence intensity of TAT peptide conjugated quantum dot (TAT-QD) in the evanescent field are attributed to the internalization of quantum dots due to a linear relationship between decreases in TAT-QDs and fluorescence density (Wang et al. 2012a). Quantitative analysis indicates that TAT-QDs exhibit rapid internalization with a 51.3% decrease in average fluorescence density after 1 h and subsequent 8% decrease in fluorescence density during the following 2 h (Wang et al. 2012a).

3.2 Particle Tracking and Photoactivated Localization Microscopy

Single-particle tracking (SPT) combines fluorescence microscopy with real-time, live cell imaging for dynamic visual tracking of individual gene delivery vectors with high spatial and temporal resolution (Dupont and Lamb 2011; Braeckmans et al. 2010a; Li et al. 2013; Ruthardt et al. 2011; Huang et al. 2011). Individual particles are visualized as unresolved dots of light; however, through the application of image processing algorithms, particle positions can be determined with sub-resolution precision, in the range of tens of nanometers, since the center of each dot can be calculated with high accuracy (Braeckmans et al. 2010a). The compilation of time-lapsed SPT movies allows for the position of particles to be determined at different time points. From this, the trajectories of each individual particle can be determined. Qualitatively, the directionality of intracellular particle movement can be traced through visual inspection. Quantitatively, the diffusivity and velocity of particles can be calculated (Dupont and Lamb 2011; Huang et al. 2011). Particles exhibit different transport modes as determined by the analysis of mean square displacements (MSD) calculated from positional data. Depending on the relationship between MSD and time, particles can be classified as undergoing: (1) normal diffusion, (2) anomalous diffusion, (3) confined diffusion, or (4) active transport (Huang et al. 2011).

SPT can provide better insight on the dynamics of different biological processes. With respect to intracellular trafficking of gene delivery vectors, SPT enables the investigation of vector localization/co-localization with cellular structures and the determination of specific barriers to intracellular delivery through direction observation in conjunction with particle kinetics data. Traditional co-localization techniques rely on either the coincidence of two fluorescence colors or close proximities of differently labeled components. As analysis is based on a single image, the occurrence of coincidental co-localization and true interaction cannot be differentiated (Braeckmans et al. 2010a). On the other hand, co-localization can be distinguished with dual-color SPT through statistically significant correlation between trajectories and kinetics of the labeled components. Thus, intracellular trafficking can be accessed through co-localizations between nucleic acid and nanoparticle vector along with interactions between the vector and other cellular components and structures (e.g., proteins, receptors, endosomes, organelles, etc.). Taking advantage of this, Vercauteren et al. (2011) implemented a trajectory-based co-localization algorithm to identify co-localizations between polyplexes and different cellular structures including clathrin-coated vesicles, recycling endosomes, early endosomes, flotillin 2, late endosomes, and endolysosomes (Vercauteren et al. 2011). Through continuous image acquisition during the first 2 h post-mRNA transfection in retinal pigment epithelium cells, the progressive transition from early endosome to late endosome and endolysosome is observed subsequently after flotillin-mediated endocytosis. In conjunction with GFP mRNA expression analysis, evidence of GFP expression within the first hour post-transfection indicates rapid internalization and endosomal escape; however, as the majority of polyplexes remained trapped in vesicles 24 h after transfection, it is postulated that cellular autophagy/xenophagy may be involved in sequestration and degradation of polyplexes. Associations between polyplexes and microtubule-associated protein 1A/1B-light chain 3 (LC3), a cellular marker for autophagy, support this notion as co-localization is detected during the first and subsequent hours post-transfection (Vercauteren et al. 2011). Similarly, co-localization analysis between PEI/DNA polyplexes and endosomes by Suh et al. (2012) indicates active transport of polyplexes within early and late endosome, both at similar rates. The presence of polyplexes in late endosomes/endolysosomes has no effect on vesicle transport rates. However, further modifications to co-localization methodology are warranted as endosomal escape was not observed in real time (Suh et al. 2012).

From SPT analysis of polyplex particle kinetics, the process of vector internalization and subsequent intracellular trafficking can be generalized into three distinct phases: (I) endocytosis, (II) anomalous or confined diffusion, and (III) microtubule-mediated active transport (Ruthardt et al. 2011; Huang et al. 2011). The attachment of polyplexes to cell membrane initiates slow directed motion characterized as endocytosis. Effects of targeted delivery are highlighted by rapid endocytosis that significantly shortens the duration of phase I. The anomalous or confined diffusion in phase II signifies impediments to normal diffusion due to the presence of various intracellular constituents. Phase III denotes long range, microtubule-mediated active transport characterized by the continuous stop-and-go motion from sequential binding and unbinding of motor proteins (Ruthardt et al. 2011; Huang et al. 2011). The three distinct phases are also observed in magnetic vectors as demonstrated by Sauer et al. (2009a). Similar to findings from Vercauteren et al. (2011), co-localization experiments by Sauer et al. (2009a) indicate that the majority of the internalized vectors are situated within internal vesicles as intact vectors with no separation between DNA and magnetic particle after 24 h (Sauer et al. 2009b).

With continuing advances to imaging techniques and the development of highly stable fluorescent probes, SPT can be achieved with ever-increasing time scales and image resolutions. By coupling two-photon microscopy and other advanced microscopy techniques with SPT, issues surrounding photodamage and photobleaching can be addressed. However, as majority of SPT experiments are generally conducted in two dimensions, particles are only tracked at specific focal planes (Dupont and Lamb 2011). Such limitation is resolved using 3D SPT through (1) post-processing of particle trajectories after image acquisition or (2) feedback-controlled real-time tracking. Post-processing 3D SPT involves the compilation of a series of 2D images (z stacks) at different axial positions and subsequent analysis of particle trajectories for the determination of diffusion behavior, diffusion coefficient, and transport velocities upon calculation of MSD (Dupont and Lamb 2011). As such, a high number of particles, within the preselected region of interest, can be simultaneously tracked. Despite the potential for multiple particle tracking, the acquisition of numerous 2D images requires extensive scanning that greatly limits time resolution (Dupont and Lamb 2011; Ruthardt et al. 2011). In feedback-controlled real-time tracking, particle positions are acquired in real time to provide feedback such that the particle of interest will remain in focus throughout the tracking period. Using this technique, high temporal resolutions can be achieved without the need for preselection of specific regions of interests (Dupont and Lamb 2011). An example of feedback-controlled real-time tracking is orbital tracking where a confocal laser beam orbits around the particle of interest. The exact position and movement of the particle within the orbit is determined depending on the fluorescence intensity distribution during each scan; any deviation in fluorescence intensity signifies a change in particle position away from the center of the orbit and provides information relating to directionality and velocity of particle movement (Ruthardt et al. 2011). Using a feedback loop, the laser is reoriented around the new position of the particle, and particle positions are recorded to ascertain particle trajectories in real time. Orbital tracking boasts several advantages including (1) high spatial resolution, (2) low phototoxicity through targeted illumination of the particle of interest, and (3) high accuracy in live cells with inherent high background fluorescence intensities (Dupont and Lamb 2011). Orbital tracking is not applicable for tracking rare events as it is restricted to tracking a limited number of particles.

SPT has been implemented in various other novel applications including the sizing of nanoparticle vectors in biological fluids. Dynamic light scattering (DLS) is regarded as an effective technique for sizing nanoparticles in simple solvents; however, DLS measurements are only accurate when nanoparticles are dispersed in the absence of other scattering components. Such complications contribute to the lack of accurate studies analyzing size and aggregation properties of nanoparticle

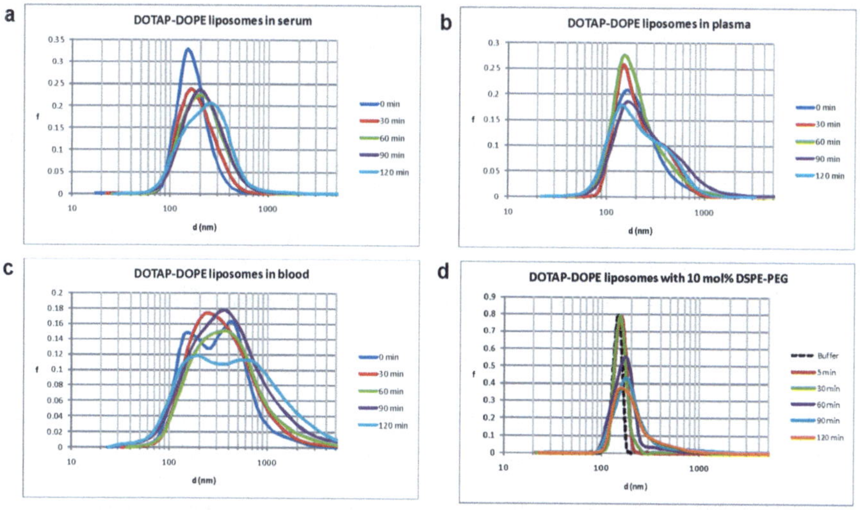

Fig. 3.7 Time-lapsed in vitro fPST size analysis of non-PEGylated DOTAP/DOPE liposomes in (**a**) serum, (**b**) plasma, (**c**) whole blood, and (**d**) PEGylated DOTAP/DOPE liposomes in whole blood. Adapted from Braeckmans et al. (2010b)

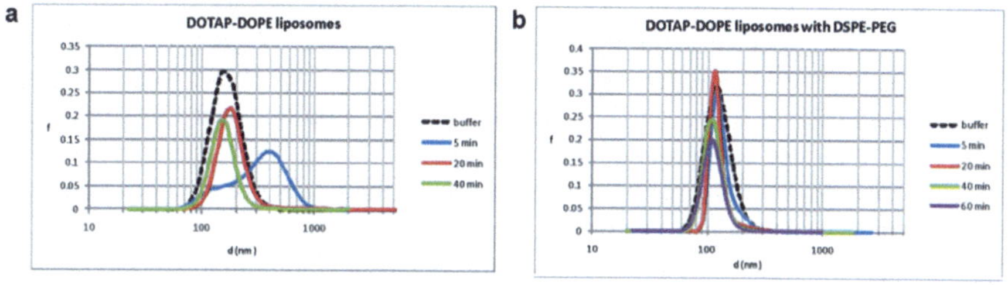

Fig. 3.8 Time-lapsed in vivo fPST size analysis of (**a**) non-PEGylated DOTAP/DOPE liposomes and (**b**) PEGylated DOTAP/DOPE liposomes in rats after intravenous injection and subsequent blood collection. Adapted from Braeckmans et al. (2010b)

vectors in biological fluids. Using fluorescence SPT (fSPT), in combination with maximum entropy analysis, Braeckmans et al. demonstrate the in vitro (Fig. 3.7) and in vivo (Fig. 3.8) analysis of nanoparticle size in undiluted whole blood in real time (Braeckmans et al. 2010b). In vitro analysis of fluorescently labeled cationic liposomes, comprised of a mixture of cationic lipid 1,2-dioleyl-3-trimethylammonium-propane (DOTAP) and 1,2-dioleoyl-sn-glycero-3-phosphoethanolamine (DOPE), indicates more pronounced aggregation over time in whole blood when compared to serum or plasma; this suggests that the in vitro assessment of size and aggregation properties in serum does not accurately reflect the in vivo conditions after intravenous injection (Braeckmans et al. 2010b). The effects of PEG coating on reducing aggregation are confirmed as PEGylation with 1,2-distearoyl-sn-glycero-3-phosphoethanolamine *N*-[methoxy(polyethylene glycol)-2000] (DSPE-PEG) prevented aggregation and resulted in smaller nanoparticle sizes. Under in vivo conditions, by means of intravenous injection into rats, time-lapsed fSPT analysis indicates rapid removal of large aggregates from circulation. The capabilities of fSPT for sizing in biological fluids permit effective characterization of nanoparticle vectors under in vivo conditions; this contributes to the development of new gene delivery vectors with improved pharmacokinetic profiles.

3.3 Quantum Dots and Fluorescence Resonance Energy Transfer

Quantum dots (QDs) are semiconductor nanocrystals composed of a core cluster of atoms of varying compositions, typically arranged in binary or ternary compound configurations with sizes ranging from 1 to 20 nm (Zhang and Wang 2012). QDs exhibit photoluminescence with their respective optical properties varying in accordance to their composition, particle size, size distribution, and surface chemistry (Resch-Genger et al. 2008). In terms of composition, QDs are commonly comprised of a CdSe or CdTe core as they can generate particles possessing the desired optical properties. Depending on particle size, emission wavelengths of QDs can be finely tuned across the visible light spectrum, while variations to QD size distribution govern the width of the emission peak (Zhang and Wang 2012; Resch-Genger et al. 2008; Barroso 2011). As QDs possess broad absorption spectra, a wide range of excitation wavelengths are suitable to enhance separation between QD excitation/emission and reduce background scattering (Barroso 2011). With respect to cellular imaging, QDs possess distinct properties that make them superior over conventional fluorescent dyes including (1) broad absorption and tunable narrow symmetric emission, (2) high quantum yields conferring enhanced photoluminescence, (3) longer fluorescence lifetimes, and (4) excellent thermal and photochemical stability contributing to strong resistance against photobleaching (Zhang and Wang 2012; Resch-Genger et al. 2008; Barroso 2011; Aalinkeel et al. 2012; Wu et al. 2012; Kuo et al. 2010; Grigsby et al. 2012; Pierobon and Cappello 2012). The broad absorption spectra of QDs, capable of extending far into UV range, permit excitation of multiple QDs using a single light source. Tunable emission spectra permit multiplexing, while longer fluorescence lifetimes and high photostability permit extended imaging (Barroso 2011; Grigsby et al. 2012; Pierobon and Cappello 2012; Brandén et al. 1999); both features are critical to co-localization and time-lapsed live-cell imaging studies. In addition, tunable emission at NIR range permits the in vivo applications of QDs in conjunction with two-photon microscopy (Altınoğlu and Adair 2010; Hutter and Maysinger 2011; Sen et al. 2008).

Common QD assembly involves the CdSe or CdTe core encased in a ZnS nanoshell to increase core stability and limit Cd-associated cytotoxicity while enhancing luminescence by reducing surface defects for improved quantum yields and narrowing emission spectrum (Zhang and Wang 2012; Resch-Genger et al. 2008; Aalinkeel et al. 2012). Thickness of the nanoshell must be carefully adjusted in order to offer adequate protection against Cd leakage without compromising QD colloidal stability and optical properties (Aalinkeel et al. 2012). Additional surface coatings are required for the generation of water dispersible QDs followed by subsequent conjugation via biotin-avidin interactions. In overcoming cytotoxicity associated with Cd-, Te-, and Se-based QDs, the development of I-III-VI_2-based QDs, such as $AgInS_2$, $CuInS_2$, and ZnS-$AgInS_2$ (ZAIS) QDs, serves to be the next step in the generation of nontoxic QDs; however, the synthesis of such QDs encompasses certain shortcomings including high reaction temperature, inadequate control over growth rates, long reaction times, complex procedures, and complications associated with high-throughput synthesis (Subramaniam et al. 2012). Addressing the above shortcomings, Subramaniam et al. have recently developed a novel sonochemical approach for high-throughput synthesis of a biocompatible library of ZAIS QDs (Subramaniam et al. 2012). The application of ultrasound irradiation in sonochemical synthesis holds several advantages including rapid reaction rates in controllable conditions, generation of QDs with uniform shape and narrow size distributions, and high-purity synthesis at ambient temperatures. In contrast to size dependence exhibited in CdSe and CdTe QDs, the optical properties of ZAIS QDs are composition dependent as emission wavelengths are controlled by varying the concentrations of each precursor element; this offers the potential for QD emissions at near UV range, a property unattainable with size dependent CdSe/CdTe QDs (Subramaniam et al. 2012). Characterization of these ZAIS QDs has demonstrated comparable quantum yields to CdSe/CdTe QDs, high stability at ambient temperatures, significantly improved biocompatibility at high concentrations, strong fluorescence inside cells, and successful siRNA delivery upon conjugation to siRNA via PEI (Subramaniam et al. 2012).

The enhanced fluorescence and photostability of QDs permits long-term, real-time imaging of delivery vectors, at nanoscale resolutions, throughout the gene delivery process from membrane binding and cellular uptake to intracellular trafficking and endosomal escape (Barroso 2011). Through the application of QDs and fluorescence microscopy, Hess et al. assessed cellular binding, surface motion, and internalization of polyarginine (PA) and polyethylenimine (PEI)-based polyplexes (Hess et al. 2007). Disruption of proteoglycans was shown to prevent stable binding and internalization of PA-QDs and PEI-QDs without affecting cell surface motion, while cholesterol sequestration was shown to inhibit both cell surface motion and internalization of PA-QDs/PEI-QDs (Hess et al. 2007). In a study assessing the intracellular trafficking of polyplexes in prostate cancer cells (PC3) and prostate cancer cells stably expressing prostate-specific membrane antigen (PSMA) (PC3-PSMA), Barua and Rege demonstrated enhanced transgene expression through the administration of histone deacetylase (HDAC) inhibitors as mediators of intracellular trafficking (Barua and Rege 2010). High co-localization of QDs and polyplexes at the perinuclear region of PC3-PSMA cells denoted microtubule-mediated transport and sequestration of QD/polyplex aggregates at the perinuclear recycling compartment (PNRC) and microtubule-organizing center (MTOC); this was in contrast to dispersed distribution of QDs/polyplexes through the cytoplasm of PC3 cells (Barua and Rege 2010). Correlations between high co-localization QDs/polyplex aggregates at PNRC/

MTOC and a relative decrease in transgene expression prompted speculations of diminished endosomal escape due to aggregation of polyplexes at a confined location within the perinuclear region. The administration of tubacin, a histone deacetylase 6 (HDAC6) inhibitor, and trichostatin A, a class I and II HDAC inhibitor, significantly enhanced transgene expression as inhibition of cytoplasmic HDACs abolished polyplex localization at PNRC/MTOC and resulted in a more uniform distribution of polyplexes(Barua and Rege 2010). Through coupling of CdTe/CdS core/shell QD with plasmid expressing the herpes simplex virus thymidine kinase (HSV-TK) gene, Shao et al. illustrated intracellular trafficking of plasmid TK and subsequent antitumoral effects after the administration of ganciclovir (GCV) for up to 96 h post-transfection (Shao et al. 2012). The strong luminescence of QDs was demonstrated to effectively trace GCV-mediated apoptosis in QD-TK-transfected HeLa cells (Shao et al. 2012).

Application of gold nanoparticle quenchers in tandem with QDs had been demonstrated to improve the signal-to-noise ratio of QD fluorescence. By adding similar-sized gold nanoparticles prior to cell imaging, Kuo et al. effectively discriminated internalized QDs from extracellular QDs as gold nanoparticles completely quenched the fluorescence of membrane bound QDs; high quenching efficiency of gold nanoparticles enhances QD imaging by reducing background QD fluorescence without inducing significant cytotoxicity (Kuo et al. 2010). Endocytosed QDs were seen to be distributed within the whole cell, and the incorporation of nuclear localization signal (NLS) to QDs permitted nuclear entry (Kuo et al. 2010).

In addition to cellular imaging, QDs have been applied for cell tracking after gene delivery. Lee et al. (2010a) developed a nanoparticle gene delivery vector comprised of QD-loaded hydrophobic poly(D,L-lactic-co-glycolic acid) (PLGA) core and positively charged glycol chitosan shell for transdermal gene delivery into Langerhans cells by gene gun bombardment; the incorporation of QDs served to track Langerhans cell migration after in vivo transdermal delivery in a mouse model. QD-labeled Langerhans cells were observed in the draining lymph node 6 h post-transfection, denoting Langerhans cell migration upon gene gun bombardment; concurrently, detectable and increasing levels of EGFP transgene expression in the lymph nodes were observed at 12 and 18 h post-transfection (Lee et al. 2010a).

With respect to QD imaging of gene delivery, QDs have been commonly used for Förster resonance energy transfer (FRET). FRET is a non-radiative energy transfer process involving dipole-dipole interactions that is highly dependent on the distances, usually <10 nm, between the energy donor and acceptor, and high FRET efficiencies require significant spectral overlap between donor emission and acceptor excitation (Zhang and Wang 2012; Barroso 2011; Grigsby et al. 2012). Optical properties of QDs offer several advantages as donors in QD-FRET: (1) broad excitation spectra and potential excitation at UV range prevent direct excitation of acceptors; (2) tunable narrow symmetric emission permits sufficient spectral overlap with acceptor excitation; and (3) large surface area enables the development of multivalent QD capable of multiple acceptor interactions for increased acceptor to donor ratio and enhanced FRET efficiencies (Zhang and Wang 2012). QD-FRET detection sensitivity is often negatively affected by QD size and surface coating;

however, such limitation may be partially compensated by improved FRET efficiencies through multiple acceptor interactions. Although QDs are excellent FRET donors, they are poor acceptors due to their extended fluorescence lifetime (Grigsby et al. 2012).

Unlike co-localization studies, QD-FRET provides critical information regarding molecular associations/interactions and dissociations. As such, QD-FRET is often employed for the assessment of complex assembly, unpacking, endosomal escape, and pDNA release. Through QD-FRET and image-based quantification, Chen et al. assessed and compared the intracellular unpacking and compartmental localization of complexed/released QD-pDNA upon gene delivery with Cy5-labeled chitosan, PEI, and polyphosphoramidate (PPA)-based polyplexes (Chen et al. 2008). Upon analysis of unpacking kinetics of the three polyplexes, rapid unpacking and pDNA release were observed with PEI polyplexes at 2 h post-transfection. In contrast, PPA polyplexes were the most stable of the three with moderate polyplex dissociation between 2 and 4 h post-transfection. The authors noted compartmental differences in unpacking as PEI polyplexes exhibited equal unpacking rates in endo/lysosomes and cytosol while PPA polyplexes predominantly unpacked within the cytosol (Chen et al. 2008). In relation to transfection efficiency, chitosan polyplexes exhibited lowered transfection efficiency due to less efficient cellular uptake, ineffective endosomal escape contributing to pDNA degradation within endo/lysosomes, and limited transport into the nucleus with majority of the released pDNA remaining in the cytosol. Although PPA polyplexes offered better cellular uptake and protection within endo/lysosomes along with more efficient transport to the nucleus, such highly stable polyplexes prevented efficient pDNA release resulting in reduced transfection efficiency (Chen et al. 2008). Overall, PEI polyplexes exhibited higher transfection efficiencies due to high cellular uptake, effective endosomal escape, efficient transport, and rapid unpacking for better delivery of free pDNA into the nucleus. Using a similar approach, Ulasov et al. found a statistically positive correlation between cellular uptake and transfection efficiency and a negative correlation between unpacking within endo/lysosomal compartments and transfection efficiency upon the delivery of PEG-TAT functionalized PEI polyplexes at varying component ratios (Ulasov et al. 2011).

In a more intricate study, Shaheen et al. demonstrated controlled unpacking of polyplexes within specific nuclear sub-domains through modulation of a biocleavable polyrotaxane synthetic polycation possessing dimethylaminoethyl-modified α-cyclodextrin, a disulfide-linked PEG chain, and a rhodamine label as FRET acceptor (DMAE-ss-PRX) (Shaheen et al. 2011). Subsequent complexation with CdSe/ZNS QD-labeled pDNA generated FRET-detectable polyplexes that are then encapsulated in a tetra-lamellar multifunctional envelope-type nanodevice (T-MEND) for nuclear delivery. T-MEND consisted of two endosome-fusogenic outer envelopes and two nuclear membrane-fusogenic inner envelopes for bypassing endosomal and nuclear membranes (Shaheen et al. 2011). Polyplexes of two polycations, 16DMAE-ss-PRX and 46DMAE-ss-PRX, each composed of 16 and 46 DMAE units, respectively, were assessed for transgene expression after T-MEND-mediated nuclear delivery. 46DMAE-ss-PRX was shown to elicit significantly higher transgene expression over 16DMAE-ss-PRX; this was due to higher nuclear delivery in addition to a higher fraction of decondensed pDNA and a greater distribution of pDNA within the transcriptionally active euchromatin (Shaheen et al. 2011). Further analysis in the distribution of condensed and decondensed pDNA in subnuclear domains indicated a higher amount of decondensed pDNA present in heterochromatin, in conjunction with lower amounts of decondensed pDNA present in euchromatin, for 16DMAE-ss-PRX polyplexes. In contrast, majority of decondensed pDNA were situated in euchromatin for 46DMAE-ss-PRX polyplexes (Shaheen et al. 2011). Differences in the localization of decondensed pDNA were speculated to be associated with interactions between polyplexes and histone-rich heterochromatin; looser complexes formed with 16DMAE-ss-PRX were more likely to interact with histones leading to decondensation, whereas 46DMAE-ss-PRX posed greater resistance to decondensation within the heterochromatin.

With respect to siRNA delivery, Lee et al. examined the influences of the cell-penetrating peptide, protein transduction domain (Hph-1) peptide, on cellular uptake and intracellular delivery through QD-FRET from siRNA-Cy5/QD-PEI (Lee et al. 2010b). Through imaging analysis, the incorporation of Hph-1 had shown more rapid and efficient cell uptake and unpacking as polyplex accumulation at the perinuclear region along with great reduction in FRET intensity, indicating polyplex unpacking, occurred after 30 min post-transfection (Lee et al. 2010b). In addition to image analysis, flow cytometry was applied for high-throughput quantification of unpacking kinetics; initial unpacking kinetic rate was drastically faster with Hph-1 as 29.7% FRET reduction, in contrast to just 5.4% for polyplexes without Hph-1, which was observed 1 h post-transfection. Despite differences in initial unpacking kinetics, majority of the polyplexes were unpacked within 5 h, and there was no difference in gene silencing after 24 h. Of note, QD-PEI polyplexes exhibited greater gene silencing than stand-alone PEI polyplexes (Lee et al. 2010b). A similar study by Wu et al., involving the delivery of Cy5-labeled oligodeoxynucleotide, was conducted to compare the cellular uptake and unpacking of lipoplexes and polyplexes (Wu et al. 2011a). Initial cellular uptake of lipoplexes, at 0.5 and 1 h post-transfection, was shown to be less significant when compared to polyplexes; this was in conjunction with a higher dissociation rate and unpacking by polyplexes. Despite such differences, the intracellular fate of both lipoplexes and polyplexes was similar as comparable knockdown effect (59.1% and 70.2% for lipoplexes and polyplexes respectively) was observed (Wu et al. 2011b).

In addition to intermolecular FRET, where pDNA is labeled with a QD donor or an acceptor and the condensing agent is labeled with the counterpart, intramolecular FRET achieved by labeling pDNA with both QD donor and acceptor can provide information regarding changes in pDNA conformation, degree of condensation, and complex stability (Biju et al. 2012). As pDNA must be labeled with definite QD donor and acceptor at specific locations beyond Förster distance, such labeling becomes meticulous and laborious. Alternatively, Biju et al. ensured efficient pDNA labeling by randomly labeling a defined number of QD donors with excess acceptors followed by FRET-mediated photodecomposition to remove acceptors situated within Förster distance of QD donors (Biju et al. 2012). Taking advantage of high QD photostability, selective photobleaching of acceptors at Förster distance of QD donors prevented intrinsic FRET prior to complexation and permitted high sensitivity in detecting pDNA condensation/decondensation upon complexation. pDNA condensation with protamine resulted in strong FRET emission and decrease in QD fluorescence, while decondensation with heparan sulfate and glutathione significantly decreased FRET efficiency (Biju et al. 2012).

By applying QD multiplexing, Chen et al. simultaneously assessed pDNA decondensation/release and pDNA stability through stepwise QD-FRET involving QD and nuclear dye (ND) (BOBO-3 or Cy3)-labeled pDNA and Cy5-labeled polymer (chitosan or PEG-b-PPA) (Chen et al. 2009b). ND dye acted as both relay donor/acceptor and an indicator of DNA integrity; QD donor drives energy transfer to intermediate ND which serves as a relay donor to Cy5 final acceptor. Successful complexation was indicated by effective energy transfer to Cy5 via ND intermediate (both ND and Cy5 are "on"). Subsequent unpacking and pDNA release were denoted by decrease in FRET efficiency between Cy5 and ND (ND "on" but Cy5 "off"). pDNA degradation was denoted by decreasing FRET efficiency between QD donor and intermediate ND due to the diffusion of DNA oligomers away from QD donors as pDNA became progressively degraded (both ND and Cy5 are "off") (Chen et al. 2009b). Upon analysis of intracellular unpacking and degradation for both polyplexes, chitosan-based polyplexes remained stable and began to unpack at 4 h post-transfection into HEK293 cells; in contrast, greater degrees of unpacking were observed with PEG-b-PPA-based polyplexes along with some pDNA degradation due to the rapid release (Chen et al. 2009b). Chitosan effectively protected pDNA, but slow unpacking kinetics hampered pDNA release rendering lower transfection efficiency when compared to PEG-b-PPA polyplexes; this was in spite of higher pDNA degradation upon unpacking from PEG-b-PPA polyplexes.

3.4 Cellular MRI and PET Labeling

The previously described fluorescence microscopy techniques enable imaging of the gene delivery process at cellular and subcellular levels; however, they are unable to provide critical information regarding the in vivo biodistribution and pharmacokinetics of the gene delivery vectors with respect to the various routes of administration. Such assessment of in vivo biodistribution and pharmacokinetics is achieved through magnetic resonance imaging (MRI) and positron-emission tomography (PET). MRI is a medical imaging technique implemented to visualize various internal structures and metabolic processes in humans and animals. MRI is based on the net magnetization effect that arises from the alignment of unpaired atomic nuclear spins when subjected to a strong magnetic field; application of a radiofrequency pulse perturbs the alignment of the spins, and subsequent relaxation back to net magnetization equilibrium state is recorded as a change in electromagnetic flux that corresponds to a MRI signal (Waerzeggers et al. 2009; Herranz et al.

2011). Images generated by MRI are generally based on contrasts from alterations to spin density or relaxation times. Contrast from spin density is dependent on the differences in the amount of available nuclear spins and is proportional to water content. Relaxation time, longitudinal relaxation time (T_1) and transverse relaxation time (T_2), refers to the time required for nuclear magnetization to return to thermal equilibrium after perturbations by a radiofrequency pulse; they are tissue dependent and are modifiable by specific contrast agents (Herranz et al. 2011). All contrast agents shorten both T_1 and T_2; however, contrast agents that significantly shorten T_1 generates increased signal intensity (positive contrast), while those that significantly shorten T_2 generates decreased signal intensity (negative contrast) (Caravan 2006). Gadolinium (Gd) chelates and iron oxide nanoparticles are examples of two commonly used MRI contrast agents. Paramagnetic gadolinium chelates are T_1 agents, while superparamagnetic iron oxide nanoparticles (SPIONs) are T_2 agents.

MRI offers several advantages that enable effective in vivo imaging of the gene delivery process. High 3D spatial resolution, along with excellent penetration depth and whole body coverage, enables noninvasive visualization over an extended period; this permits quantitative monitoring of cell/vector trafficking in vivo (Liu et al. 2011). In addition, physiological and anatomical data are simultaneously gathered through MRI (Kang and Chung 2008), providing relevant information pertinent to assessing the biodistribution and pharmacokinetics of the administered vector. Lastly, MRI is safer than PET and computed tomography (CT) as it does not involve exposure to ionizing radiation. However, along with its inherent advantages, MRI is disadvantaged by poor sensitivity, high associated costs, and the requirement of highly trained personnel for operation and image analysis (Waerzeggers et al. 2009; Liu et al. 2011; Kang and Chung 2008).

SPIONs have been increasingly utilized in monitoring the in vivo spatiotemporal migration of cells and vectors through MRI because it possesses several inherent properties. SPIONs exhibit low toxicity as they are readily degraded, metabolized, and sequestered into Fe pools for various metabolic pathways including hemoglobin synthesis. However, as toxicity levels are dose dependent, significantly high doses of SPIONs may potentially induce oxidative stress due to excessive reactive oxygen species (ROS) from (1) nanoparticle-induced ROS production, (2) leaching of SPIONS, and (3) alteration/induction of different signaling pathways (Liu et al. 2013). SPIONs have large surface areas suitable for effective functional modifications. As an imaging and contrast agent, SPIONs provide the most change in MRI signal per unit of metal and are detectable by histology, through Perl's iron staining, and electron microscopy (Liu et al. 2011). Because SPIONs are concurrently used as a gene delivery agent in magnetofection, the biodistribution and pharmacokinetics of such magnetic vectors can be elucidated without any additional modifications or functionalizations. The capabilities of gene delivery and in vivo molecular imaging signify the potential for SPIONs in theranostic applications. The major drawback associated with SPIONs, and other T_2 agents, is the generation of darkened magnetic resonance images due to their negative contrast effects.

The dual functionality of SPIONS, as gene delivery and MRI agents, is exemplified in a study by Wang et al., where a SPION-loaded micelle, composed of self-assembling poly(D,L-lactide) (PLA) and monomethoxy polyethylene glycol (mPEG) block copolymers coated with chitosan/PEI, has been shown to yield higher transfection efficiency in vitro than Lipofectamine or PEI alone while having minimal hindrance to T_2 relaxivity from the DNA/polymer coating (Wang et al. 2012b). By developing tunable iron oxide mesoporous silica nanospheres, Zhao et al. controlled the corresponding doses of contrast agent and nucleic acid for the potential of a safe and effective theranostic agent. The amount of iron oxide, situated at the core of the mesoporous silica shell, can be modulated to meet the requirements for MRI in different applications, while the extent of surface positive charge functionalization can be controlled to regulate the level of nucleic acid complexation. In vitro analysis confirms the presence of negative contrast and capacity for gene delivery through reporter gene expression of EGFP (Zhao et al. 2012).

Negative contrast effects from T_2 agents often encounter issues in discriminating adjacent regions with reduced signal intensities. T_1 agents like Gd-chelates are predominantly used in contrast-enhanced clinical exams as these agents produce bright spots on MRI (Caravan 2006). Using lipid-peptide vectors comprised of cationic DOTAP/DOPE liposomes with 15% Gd-chelated lipid, receptor-targeted DNA-binding peptide, and plasmid DNA, Writer et al. demonstrated enhanced MRI contrast in vitro over the first 4 h of transfection due to increasing accessibility of Gd^{+3} to water as a result of sequential binding, internalization, endosomal escape, and disassembly of vector complexes in cells (Writer et al. 2012). The gradual decrease in MRI contrast, from peak contrast at 4 h post-transfection, was attributed to the dilution of Gd^{+3} from cell division. Such conclusions arise from time-course studies where the kinetics of vector delivery and trafficking corresponded well with the kinetics of MRI contrast enhancement. The authors validated the potential for real-time monitoring of gene delivery in the brain as the in vivo delivery of the vectors to rat brains produces detectable MRI signals at 6, 24, and 48 h post administration; the focal distribution of MRI signals is consistent with intracellular distribution of Gd^{+3} upon vector delivery (Writer et al. 2012). Using a similar vector possessing neurotensin-targeting peptides, Kenny et al. visualized the presence of vectors 4 h post-intratumoral injection into mouse models harboring tumors derived from neuroblastoma cells (Kenny et al. 2012). The enhanced contrast permitted in vivo monitoring of vector localization by MRI for extended periods up to 24 h (Kenny et al. 2012). Thus, using MRI, the in vivo distribution and retention of Gd-labeled vectors can be assessed in real time. Aside from Gd-chelates, manganese-based nanoparticles have been recently assessed to be a suitable T_1 contrast agent for MRI. Using Herceptin functionalized 3,4-dihydroxy-L-phenylalanine (DOPA)-PEI-coated hollow manganese oxide nanoparticles, Bae et al. demonstrated targeted detection of cancer cells in T_1-weighted MRI (Bae et al. 2011). Successful delivery of vascular endothelial growth factor (VEGF) siRNA permits effective cancer cell-specific silencing. From this, the visualization of tumor lesions and the monitoring of therapeutic effects associated with targeted gene delivery in tumors can be simultaneously achieved (Bae et al. 2011).

Despite the clinical use of Gd complexes as positive contrasts, the potential risk of toxic Gd^{+3} ion release, low relaxivity, and short half-lives are three disadvantages associated with their use in MRI (Meng Meng Lin et al. 2008). Recent developments of SPIONs as high-resolution T_1 agents (Tromsdorf et al. 2009; Kim et al. 2011) combine the benefits of high relaxivity, low toxicity, and positive contrast; this has serious implications for the use of SPIONs in numerous applications including in vivo monitoring of gene delivery process by MRI.

The limited spatial resolution of MRI inhibits the assessment of intracellular gene delivery processes. Often MRI is coupled with fluorescence/luminescence imaging to circumvent issues of low spatial resolution and allow subcellular visualization of vectors upon internalization and intracellular trafficking. Various vector formulations, developed by numerous research groups, have been shown to exhibit dual modality imaging capabilities. Lee et al. have synthesized a magnetic gene delivery vector composed of manganese-doped magnetism-engineered iron oxide (MnMEIO) nanoparticles bound to PEG-RGD and siGFP-Cy5 by a cleavable disulfide linker (Lee et al. 2009). The arginine-glycine-aspartic acid (RGD) serves as a targeting moiety for targeted delivery in cancer cells, while the Cy5 fluorescent dye permits tracking of siRNA intracellular trafficking (Lee et al. 2009). In addition, subsequent gene-silencing effects of siGFP can be fluorescently determined by the reduction in GFP fluorescence after successful delivery of siRNA. In vitro cellular MRI confirms the effects of targeted delivery as contrast intensity increases with increasing concentrations of the nanoparticle vector; such MR contrast is not observed in receptor-deficient cells (Lee et al. 2009). Optical imaging by confocal fluorescence microscopy permits the assessment of targeted delivery and intracellular trafficking of siRNA. The presence of cleavable disulfide linkers constitutes significant dose-dependent gene-silencing effects as it enables siRNA dissociation upon internalization (Lee et al. 2009). Namgung et al. developed a magnetic gene delivery vector with dual modality imaging capabilities using silicon nanotubes (SNT) encasing with iron oxide and quantum dots; the SNTs are surface functionalized by branched PEI (BPEI) for effective complexation with plasmid DNA (BPEI-SNT) (Fig. 3.9) (Namgung et al. 2011). Silicon nanotubes are specifically utilized in this manner as the two distinct domains, inner void and outer surface, can be functionalized for the generation of a multifunctional vector. In addition, the transparency of silicon nanotubes to UV and visible light permit the incorporation of fluorescent dyes for optical imaging of intracellular delivery (Namgung et al. 2011). Co-localization imaging of endocytic uptake is demonstrated through concurrent labeling of cells with Lysotracker red. In vitro cellular MRI supports the potential of BPEI-SNT as a MRI agent as the enhanced MRI contrast was greater than with SNT alone. Although transfection efficiency of BPEI-SNT is lower than $BPEI_{25K}$, the associated cytotoxicity of BPEI-SNT is also lower as indicated by significantly greater cell viability when compared to $BPEI_{25K}$ (Namgung et al. 2011). Exploiting the inherent properties of lanthanide metals has resulted in developing vectors capable of detection by luminescence and MRI. Luminescent lanthanide chelates, like europium (Eu) chelates, are small hydrophilic structures with long luminescence lifetimes and high photostability (Namgung et al. 2011). In conjunction with Gd-chelates acting as T_1 agents, the combination of the two chelates enables dual modality imaging as demonstrated by Bryson et al. (2009), where Eu- and Gd-chelates are bound to a cationic polymer for the generation of

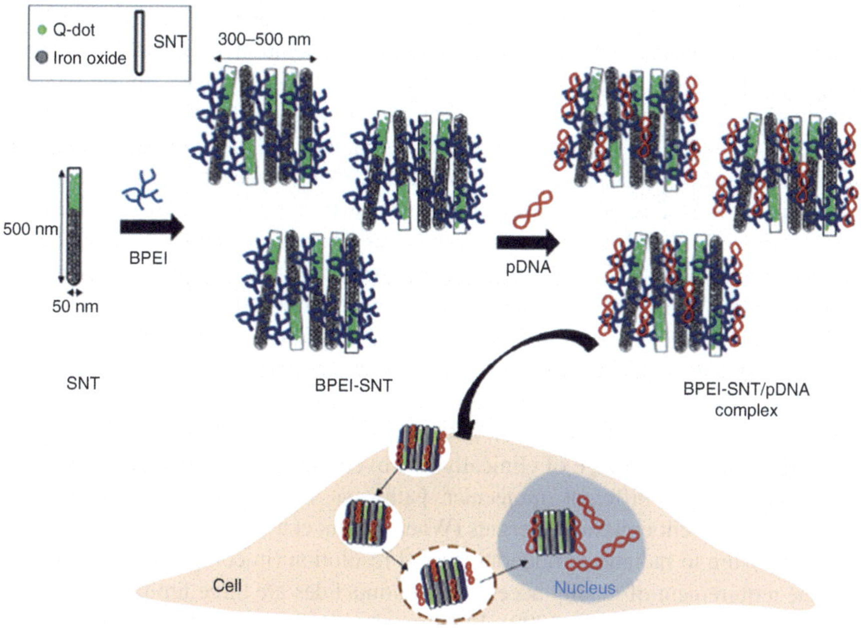

Fig. 3.9 Schematic illustration of BPEI-SNT and its subsequent intracellular delivery into cells. Adapted from Namgung et al. (2011)

Fig. 3.10 Schematic of polymer beacons comprising of a luminescent or MRI domain bound to a cationic domain possessing repeating oligoethyleneamines for DNA binding and compaction. Adapted from Bryson et al. (2009)

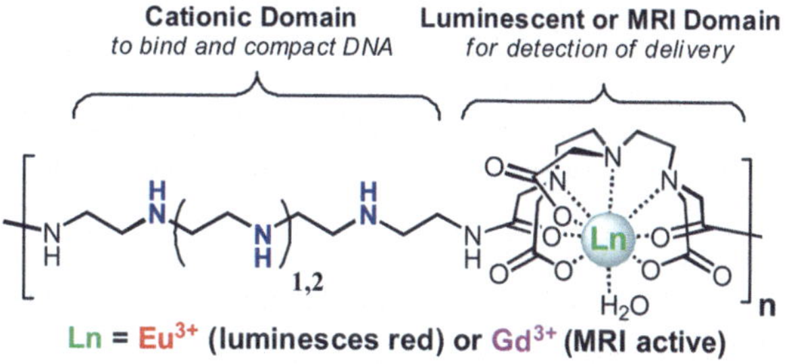

polymer beacons (Fig. 3.10). These beacons effectively complex DNA, and the subsequent delivery of the nanoparticle vector is achieved with minimal toxicity; co-localization imaging with Eu-chelated polymer beacons and fluorescently labeled DNA has been illustrated (Bryson et al. 2009). The application of near-infrared (NIR) fluorescence imaging enables in vivo tracking that corroborates well with data from MRI. Such features are demonstrated by Medarova et al., using a siRNA vector composed of a dextran-coated superparamagnetic nanoparticles functionalized with siGFP and labeled with near-infrared Cy5.5 dye (Medarova et al. 2007). In vivo optical imaging is used to both corroborate with MRI analysis of successful vector delivery and evaluate gene-silencing effect in tumors deriving from gliosarcoma cells constitutively expressing GFP (Medarova et al. 2007).

Positron-emission tomography (PET) is a form of nuclear medical imaging utilizing positron-emitting radionuclides for the generation of three-dimensional images of the body. Positron emission from radionuclides leads to subsequent annihilation events upon collision between the emitted positrons and nearby electrons; this results in the release of high-energy gamma photons in opposite direction at 180 degrees from each other (Basu et al. 2011). Coincidence detection of multiple photon pairs by PET detectors results in image construction determining the quantity and location of emitted positrons within the body. Hence, radiolabeling gene delivery vectors enable in vivo tracking of vector biodistribution and pharmacokinetics. PET imaging is commonly conducted in tandem with MRI or computed tomography (CT) for simultaneous acquisition of anatomical and functional information.

Advantages to PET imaging include (1) high sensitivity, (2) quantifiable measurements, (3) unlimited depth penetration, (4) excellent signal-to-noise ratio, and (5) long half-life (Waerzeggers et al. 2009; Kang and Chung 2008). In addition, the availability of a broad range of clinically approved imaging agents permits simultaneous imaging of different molecular pathways using different radionuclides possessing different emission energies (Waerzeggers et al. 2009). However, prerequisite exposure to radiation, moderate spatial resolution (in comparison with MRI), and the requirement of readily accessible radionuclides are three limitations to PET imaging (Waerzeggers et al. 2009). PET imaging is similar to MRI in terms of the sophisticated instrumentation and highly trained technicians required for operation and image analysis.

For in vivo tracking of gene delivery, vectors must be labeled with radionuclides possessing relatively long half-lives. Of which, copper radionuclides (^{64}Cu) are suitable for PET imaging due to their long half-lives, decay characteristics, and ease of production (Li et al. 2012). ^{64}Cu has been previously used by Bartlett et al. for in vivo tracking of siRNA nanoparticle vectors comprising of 1,4,7,10-tetraazacyclododecane-1,4,7,10-tetraacetic acid (DOTA)-siLuc encapsulated in PEGylated cyclodextrin-containing polycations (Bartlett et al. 2007). The conjugation of DOTA, a metal chelator, to siLuc enables ^{64}Cu labeling, while the attachment of transferrin targeting ligand permits targeted delivery to tumor cells. Using PET/CT imaging for the analysis of tumor localization and whole-body distribution kinetics, it is shown that vector accumulation in tumor sites is predominantly driven by the enhanced permeability and retention (EPR) effect as targeted and nontargeted vectors exhibit similar blood clearance and tumor localization/accumulation; however, the effects of targeted delivery are observed through greater reduction in luciferase activity, as indicated by reduction in bioluminescence, from increased siRNA internalization to intracellular delivery (Fig. 3.11) (Bartlett et al. 2007). Through in vitro analysis, Li et al. propose the feasibility of using cyclam-based polymeric copper chelators for gene delivery and in vivo PET imaging (Li et al. 2012). The synthesis of reducible polycationic chelator polymers, through copolymerization of cyclam and *N,N′*-cystaminebisacrylamide (CBA), permits complexation with both ^{64}Cu and nucleic acid. High complexation affinity and high stability of the cyclam-copper complexes are beneficial for PET imaging; however, enhanced

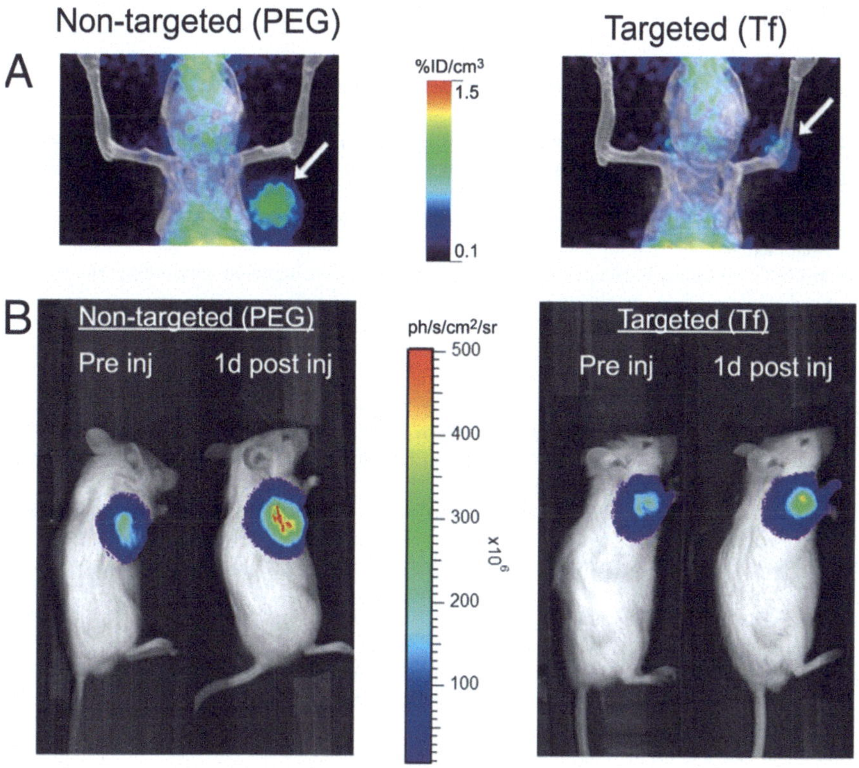

Fig. 3.11 (**a**) MicroPET/CT images highlighting tumor accumulation (white arrow) 1 day after injection of targeted and nontargeted vectors; (**b**) bioluminescence imaging pre- and 1 h postinjection indicating greater silencing effect with targeted vectors. Adapted from Bartlett et al. (2007)

polyplex stability from Cu complexation impedes effective disassembly and subsequent nucleic acid release. As a result, transfection efficiency decreases with increasing Cu complexation (Li et al. 2012).

Limited accessibility to radionuclides is a major limitation preventing PET-based in vivo tracking of gene delivery, and the development of conveniently labeled PET image-capable vectors are highly sought after. Gao et al. have demonstrated the synthesis of various ^{11}C-labeled cholesterol-based cationic lipids for gene delivery and in vivo PET imaging (Gao et al. 2010). By implementing an efficient synthesis scheme, the authors can generate PET probes with high radiochemical yields and great specific activity while shortening the overall synthesis process (Gao et al. 2010). Prospects from this study promote the generation of new PET probes suitable for in vivo tracking of gene delivery.

References

Aalinkeel R, Nair B, Reynolds JL et al (2012) Quantum rods as nanocarriers of gene therapy. Drug Deliv 19:220–231

Aaron JS, Greene AC, Kotula PG et al (2011) Advanced optical imaging reveals the dependence of particle geometry on interactions between CdSe quantum dots and immune cells. Small 7:334–341

Altınoğlu Eİ, Adair JH (2010) Near infrared imaging with nanoparticles. Wiley Interdiscip Rev Nanomed Nanobiotechnol 2:461–477

Andresen V, Alexander S, Heupel W-M et al (2009) Infrared multiphoton microscopy: subcellular-resolved deep tissue imaging. Curr Opin Biotechnol 20:54–62

Bae KH, Lee K, Kim C, Park TG (2011) Surface functionalized hollow manganese oxide nanoparticles for cancer targeted siRNA delivery and magnetic resonance imaging. Biomaterials 32:176–184

Barroso MM (2011) Quantum dots in cell biology. J Histochem Cytochem 59:237–251

Bartlett DW, Su H, Hildebrandt IJ et al (2007) Impact of tumor-specific targeting on the biodistribution and efficacy of siRNA nanoparticles measured by multimodality in vivo imaging. Proc Natl Acad Sci 104:15549–15554

Barua S, Rege K (2010) The influence of mediators of intracellular trafficking on transgene expression efficacy of polymer–plasmid DNA complexes. Biomaterials 31:5894–5902

Basu S, Kwee TC, Surti S et al (2011) Fundamentals of PET and PET/CT imaging. Ann N Y Acad Sci 1228:1–18

Bélanger E, Turcotte R, Daradich A et al (2015) Maintaining polarization in polarimetric multiphoton microscopy. J Biophoton 8:884–888

Biju V, Anas A, Akita H et al (2012) FRET from quantum dots to photodecompose undesired acceptors and report the condensation and decondensation of plasmid DNA. ACS Nano 6:3776–3788

Braeckmans K, Buyens K, Naeye B et al (2010a) Advanced fluorescence microscopy methods illuminate the transfection pathway of nucleic acid nanoparticles. J Control Release 148:69–74

Braeckmans K, Buyens K, Bouquet W et al (2010b) Sizing nanomatter in biological fluids by fluorescence single particle tracking. Nano Lett 10:4435–4442

Brandén LJ, Mohamed a J, Smith CI (1999) A peptide nucleic acid-nuclear localization signal fusion that mediates nuclear transport of DNA. Nat Biotechnol 17:784–787. https://doi.org/10.1038/11726

Brandenburg B, Zhuang X (2007) Virus trafficking - learning from single-virus tracking. Nat Rev Microbiol 5:197–208

Bryson JM, Fichter KM, Chu W-J et al (2009) Polymer beacons for luminescence and magnetic resonance imaging of DNA delivery. Proc Natl Acad Sci U S A 106:16913–16918

Caravan P (2006) Strategies for increasing the sensitivity of gadolinium based MRI contrast agents. Chem Soc Rev 35:512

Chen HH, Ho Y-P, Jiang X et al (2008) Quantitative comparison of intracellular unpacking kinetics of polyplexes by a model constructed from quantum dot-FRET. Mol Ther 16:324–332

Chen H, Wang H, Slipchenko MN et al (2009a) A multimodal platform for nonlinear optical microscopy and microspectroscopy. Opt Express 17:1282–1290. https://doi.org/10.1364/OE.17.001282

Chen HH, Ho Y-P, Jiang X et al (2009b) Simultaneous non-invasive analysis of DNA condensation and stability by two-step QD-FRET. Nano Today 4:125–134

Cheng S-H, Li F-C, Souris JS et al (2012) Visualizing dynamics of sub-hepatic distribution of nanoparticles using intravital multiphoton fluorescence microscopy. ACS Nano 6:4122–4131

Christensen DJ, Nedergaard M (2011) Two-photon in vivo imaging of cells. Pediatr Nephrol 26:1483–1489

Christensen DJ, Gottlin EB, Benson RE, Hamilton PT (2001) Phage display for target-based antibacterial drug discovery. Drug Discov Today 6:721–727

Chu S-W, Chen S-Y, Tsai T-H et al (2003) In vivo developmental biology study using noninvasive multi-harmonic generation microscopy. Opt Express 11:3093–3099. https://doi.org/10.1364/OE.11.003093

de Vito G, Bifone A, Piazza V (2012) Rotating-polarization CARS microscopy: combining chemical and molecular orientation sensitivity. Opt Express 20:29369

Dupont A, Lamb DC (2011) Nanoscale three-dimensional single particle tracking. Nanoscale 3:4532

Fritzky L, Lagunoff D (2013) Advanced methods in fluorescence microscopy. Anal Cell Pathol 36:5–17

Gao M, Wang M, Miller KD et al (2010) Facile synthesis of carbon-11-labeled cholesterol-based cationic lipids as new potential PET probes for imaging of gene delivery in cancer. Steroids 75:715–720

Grigsby CL, Ho Y-P, Leong KW (2012) Understanding nonviral nucleic acid delivery with quantum dot-FRET nanosensors. Nanomedicine 7:565–577

Hayek A, Ercelen S, Zhang X et al (2007) Conjugation of a new two-photon fluorophore to poly (ethylenimine) for gene delivery imaging. Bioconjug Chem 18:844–851

Herranz F, Almarza E, Rodríguez I et al (2011) The application of nanoparticles in gene therapy and magnetic resonance imaging. Microsc Res Tech 74:577–591. https://doi.org/10.1002/jemt.20992

Hess GT, Humphries WH, Fay NC, Payne CK (2007) Cellular binding, motion, and internalization of synthetic gene delivery polymers. Biochim Biophys Acta - Mol Cell Res 1773:1583–1588

Hoover EE, Squier JA (2013) Advances in multiphoton microscopy technology. Nat Photon 7:93–101

Huang F, Dempsey C, Chona D, Suh J (2011) Quantitative nanoparticle tracking: applications to nanomedicine. Nanomedicine 6:693–700. https://doi.org/10.2217/nnm.11.42

Hutter E, Maysinger D (2011) Gold nanoparticles and quantum dots for bioimaging. Microsc Res Tech 74:592–604

Jouvenet N, Bieniasz PD, Simon SM (2008) Imaging the biogenesis of individual HIV-1 virions in live cells. Nature 454:236–240

Kang JH, Chung J-K (2008) Molecular-genetic imaging based on reporter gene expression. J Nucl Med 49:164S–179S

Kato M, Akao M, Matsumoto-Ida M et al (2009) The targeting of cyclophilin D by RNAi as a novel cardioprotective therapy: evidence from two-photon imaging. Cardiovasc Res 83:335–344

Kenny GD, Villegas-Llerena C, Tagalakis AD et al (2012) Multifunctional receptor-targeted nanocomplexes for magnetic resonance imaging and transfection of tumours. Biomaterials 33:7241–7250

Kim BH, Lee N, Kim H et al (2011) Large-scale synthesis of uniform and extremely small-sized iron oxide nanoparticles for high-resolution T_1 magnetic resonance imaging contrast agents. J Am Chem Soc 133:12624–12631

Kobat D, Horton NG, Xu C (2011) In vivo two-photon microscopy to 1.6-mm depth in mouse cortex. J Biomed Opt 16:106014

Kuo K-W, Chen T-H, Kuo W-T et al (2010) Cell uptake and intracellular visualization using quantum dots or nuclear localization signal-modified quantum dots with gold nanoparticles as quenchers. J Nanosci Nanotechnol 10(7):4173

Lee J-H, Lee K, Moon SH et al (2009) All-in-one target-cell-specific magnetic nanoparticles for simultaneous molecular imaging and siRNA delivery. Angew Chem Int Ed 48:4174–4179

Lee P-W, Hsu S-H, Tsai J-S et al (2010a) Multifunctional core-shell polymeric nanoparticles for transdermal DNA delivery and epidermal Langerhans cells tracking. Biomaterials 31:2425–2434

Lee H, Kim I-K, Park TG (2010b) Intracellular trafficking and unpacking of siRNA/quantum dot-PEI complexes modified with and without cell penetrating peptide: confocal and flow cytometric FRET analysis. Bioconjug Chem 21:289–295

Li Q, Wu SSH, Chou KC (2009) Subdiffraction-limit two-photon fluorescence microscopy for GFP-tagged cell imaging. Biophys J 97:3224–3228
Li J, Zhu Y, Hazeldine ST et al (2012) Cyclam-based polymeric copper chelators for gene delivery and potential PET imaging. Biomacromolecules 13:3220–3227
Li W, Liu R, Wang Y et al (2013) Temporal techniques: dynamic tracking of nanomaterials in live cells. Small 9:1585–1594
Liu G, Swierczewska M, Niu G et al (2011) Molecular imaging of cell-based cancer immunotherapy. Mol Biosyst 7:993
Liu G, Gao J, Ai H, Chen X (2013) Applications and potential toxicity of magnetic iron oxide nanoparticles. Small 9:1533–1545
Martini J, Schmied K, Palmisano R et al (2007) Multifocal two-photon laser scanning microscopy combined with photo-activatable GFP for in vivo monitoring of intracellular protein dynamics in real time. J Struct Biol 158:401–409
Mattheyses AL, Axelrod D (2006) Direct measurement of the evanescent field profile produced by objective-based total internal reflection fluorescence. J Biomed Opt 11:14006
Mattheyses AL, Simon SM, Rappoport JZ (2010) Imaging with total internal reflection fluorescence microscopy for the cell biologist. J Cell Sci 123
Mattheyses AL, Atkinson CE, Simon SM (2011) Imaging single endocytic events reveals diversity in clathrin, dynamin and vesicle dynamics. Traffic 12:1394–1406
Medarova Z, Pham W, Farrar C et al (2007) In vivo imaging of siRNA delivery and silencing in tumors. Nature Publishing Group
Meng Meng Lin MM, Do Kyung Kim DK, El Haj AJ, Dobson J (2008) Development of superparamagnetic iron oxide nanoparticles (SPIONS) for translation to clinical applications. IEEE Trans Nanobiosci 7:298–305
Mojzisova H, Vermot J (2011) When multiphoton microscopy sees near infrared. Curr Opin Genet Dev 21:549–557
Namgung R, Zhang Y, Fang QL et al (2011) Multifunctional silica nanotubes for dual-modality gene delivery and MR imaging. Biomaterials 32:3042–3052
Paoli J, Smedh M, Ericson MB (2009) Multiphoton laser scanning microscopy—a novel diagnostic method for superficial skin cancers. Semin Cutan Med Surg 28:190–195
Parhamifar L, Moghimi SM (2012) Total internal reflection fluorescence (TIRF) microscopy for real-time imaging of nanoparticle-cell plasma membrane interaction. In: Nanoparticles in biology and medicine. Humana, Totowa, NJ, pp 473–482
Pierobon P, Cappello G (2012) Quantum dots to tail single bio-molecules inside living cells. Adv Drug Deliv Rev 64:167–178
Resch-Genger U, Grabolle M, Cavaliere-Jaricot S et al (2008) Quantum dots versus organic dyes as fluorescent labels. Nat Methods 5:763–775
Ruthardt N, Lamb DC, Bräuchle C (2011) Single-particle tracking as a quantitative microscopy-based approach to unravel cell entry mechanisms of viruses and pharmaceutical nanoparticles. Mol Ther 19:1199–1211. https://doi.org/10.1038/mt.2011.102
Sauer AM, de Bruin KG, Ruthardt N et al (2009a) Dynamics of magnetic lipoplexes studied by single particle tracking in living cells. J Control Release 137:136–145. https://doi.org/10.1016/j.jconrel.2009.04.003
Sauer AM, de Bruin KG, Ruthdart N, et al (2009b) CIPSM - dynamics of magnetic lipoplexes studied by single particle tracking in living cells. J Control Release 137:136.145.
Sen D, Deerinck TJ, Ellisman MH et al (2008) Quantum dots for tracking dendritic cells and priming an immune response in vitro and in vivo. PLoS One 3:e3290
Shaheen SM, Akita H, Yamashita A et al (2011) Quantitative analysis of condensation/decondensation status of pDNA in the nuclear sub-domains by QD-FRET. Nucleic Acids Res 39:e48–e48
Shao D, Zeng Q, Fan Z et al (2012) Monitoring HSV-TK/ganciclovir cancer suicide gene therapy using CdTe/CdS core/shell quantum dots. Biomaterials 33:4336–4344

Son JH, Lim CS, Han JH et al (2011) Two-photon lysotrackers for in vivo imaging. J Org Chem 76:8113–8116

Steyer JA, Almers W (1999) Tracking single secretory granules in live chromaffin cells by evanescent-field fluorescence microscopy. Biophys J 76:2262–2271

Subramaniam P, Lee SJ, Shah S et al (2012) Generation of a library of non-toxic quantum dots for cellular imaging and siRNA delivery. Adv Mater 24:4014–4019

Suh J, An Y, Tang BC et al (2012) Real-time gene delivery vector tracking in the endo-lysosomal pathway of live cells. Microsc Res Tech 75:691–697

Svoboda K, Yasuda R (2006) Principles of two-photon excitation microscopy and its applications to neuroscience. Neuron 50:823–839. https://doi.org/10.1016/j.neuron.2006.05.019

Tkaczyk ER, Tkaczyk AH (2011) Multiphoton flow cytometry strategies and applications. Cytom Part A 79A:775–788

Toomre D, Manstein DJ (2001) Lighting up the cell surface with evanescent wave microscopy. Trends Cell Biol 11:298–303

Trexler AJ, Taraska JW (2017) Two-color total internal reflection fluorescence microscopy of exocytosis in endocrine cells. Humana, New York, NY, pp 151–165

Tromsdorf UI, Bruns OT, Salmen SC et al (2009) A highly effective, nontoxic T_1 MR contrast agent based on ultrasmall PEGylated iron oxide nanoparticles. Nano Lett 9:4434–4440

Turcotte R, Rutledge DJ, Bélanger E et al (2016) Intravital assessment of myelin molecular order with polarimetric multiphoton microscopy. Sci Rep 6:31685

Ulasov AV, Khramtsov YV, Trusov GA et al (2011) Properties of PEI-based polyplex nanoparticles that correlate with their transfection efficacy. Mol Ther 19:103–112

Ustione A, Piston DW (2011) A simple introduction to multiphoton microscopy. J Microsc 243:221–226

Vercauteren D, Deschout H, Remaut K et al (2011) Dynamic colocalization microscopy to characterize intracellular trafficking of nanomedicines. ACS Nano 5:7874–7884

Waerzeggers Y, Monfared P, Viel T et al (2009) Methods to monitor gene therapy with molecular imaging. Methods 48:146–160

Wang X, Yao S, Ahn H-Y et al (2010) Folate receptor targeting silica nanoparticle probe for two-photon fluorescence bioimaging. Biomed Opt Express 1:453–462

Wang J, Zhan Y, Bao N, Lu C (2012a) Quantitative measurement of quantum dot uptake at the cell population level using microfluidic evanescent-wave-based flow cytometry. Lab Chip 12:1441

Wang C, Ravi S, Martinez GV et al (2012b) Dual-purpose magnetic micelles for MRI and gene delivery. J Control Release 163:82–92

Writer MJ, Kyrtatos PG, Bienemann AS et al (2012) Lipid peptide nanocomplexes for gene delivery and magnetic resonance imaging in the brain. J Control Release 162:340–348

Wu M, Kempaiah R, Huang P-JJ et al (2011a) Adsorption and desorption of DNA on graphene oxide studied by fluorescently labeled oligonucleotides. Langmuir 27:2731–2738. https://doi.org/10.1021/la1037926

Wu Y, Ho Y-P, Mao Y et al (2011b) Uptake and intracellular fate of multifunctional nanoparticles: a comparison between lipoplexes and polyplexes via quantum dot mediated Förster resonance energy transfer. Mol Pharm 8:1662–1668

Wu Y, Eisele K, Doroshenko M et al (2012) A quantum dot photoswitch for DNA detection, gene transfection, and live-cell imaging. Small 8:3465–3475

Zhang Y, Wang T-H (2012) Quantum dot enabled molecular sensing and diagnostics. Theranostics 2:631–654

Zhao Y, Lin L-N, Lu Y et al (2012) Synthesis of tunable theranostic Fe^3O^4@mesoporous silica nanospheres for biomedical applications. Adv Healthc Mater 1:327–331

Chapter 4
Specific Systems for Evaluation

Roderick Adrian Slavcev, Chi Hong Sum, Jesse St. Jean, Haein Huh, and Nafiseh Nafissi

Contents

Abstract Fluorescent-based visualization techniques have long been used to monitor biological activity. This chapter explores the delivery of reporter genes as a means to assay and track activity in biological systems. Bioluminescence is the production of light due to biochemical processes. By encoding genes for bioluminescence, biological processes can be visualized based on gene expression. This chapter also discusses the primary applications of bioluminescence as seen through bioluminescent imaging techniques, flow cytometry, and PCR-based methods of gene detection. These techniques are described in terms of researching gene expression, cancer therapy, and protein interactions.

Keywords Bioluminescence · Luciferase · Bioluminescence resonance energy transfer (BRET) · Fluorescence resonance energy transfer (FRET) · Photoproteins · Green fluorescent protein · Real-time PCR · Reporter genes

R. A. Slavcev (✉) · C. H. Sum · J. St. Jean · H. Huh · N. Nafissi
University of Waterloo, School of Pharmacy, Waterloo, ON, Canada
e-mail: slavcev@uwaterloo.ca

© Springer International Publishing AG, part of Springer Nature 2018

R. A. Slavcev et al. (eds.), *Nanomedicine*, Experientia Supplementum 110,
https://doi.org/10.1007/978-3-319-78259-1_4

4.1 Reporter Gene Expression

4.1.1 Fluorescence: GFP/YFP/RFP

The absorption and emission of light by a specimen is a phenomenon referred to as fluorescence, and a fluorescent microscope enables the visualization of cellular structures or processes. It is an indispensable tool in biology, with a wide range of applications in visualizing and quantifying both gene and protein expressions as well as cell-cell and molecular interactions (Liu et al. 2015). Fluorescence can be conferred to a molecule via conjugation of fluorophores, which are chemical compounds that absorb and re-emit light upon excitation at a specific wavelength. Oftentimes, they are covalently bound to macromolecules and serve as labels. Fluorophores can be broadly classified into two groups—intrinsic, which refers to fluorescence occurring naturally in macromolecules, and extrinsic, whereby fluorophore probes must be added to the system (Hofmann 2010). Proteins can comprise of amino acids that possess intrinsic fluorescence including aromatic amino acids, such as tryptophan, tyrosine, and phenylalanine, and can be differentiated upon excitation with corresponding wavelengths (Lakowicz 2006; Permyakov 1993). Other macromolecules, such as DNA and lipids, are completely devoid of intrinsic fluorescence. Fluorescence must be acquired by labeling the molecule with extrinsic fluorescent probes, typically by a non-covalent bond or chemical coupling (Hofmann 2010). This is upon three conditions of the fluorophores, as described below (Hofmann 2010):

1. The fluorophore must be tightly bound at a unique location.
2. The fluorophore should be sensitive to changing environmental conditions.
3. The fluorophore itself must not interfere with the structure and function of the macromolecule being investigated.

Fluorescent labels such as the green fluorescent protein (GFP) from the bioluminescent jellyfish *Aequorea victoria* can be tagged to proteins of interest for spatial and temporal monitoring of the labeled protein, including second-messenger cascades and intracellular signaling (van Roessel and Brand 2002). GFP (Fig. 4.1) is a 28 kDa protein (238 a.a.) characterized by excitation wavelengths of 395 nm and emission of 509 nm, which is visualized as green light (Chalfie et al. 1994). Similarly, there are other variants from corals, including red fluorescent protein (RFP) from *Discosoma striata* and yellow fluorescent protein (YFP) that have maximum emission spectra ranging from 448 to 600 nm. One of the biggest advantages of GFP is the ability of the chromophore in GFP to form spontaneously without the presence of a substrate or cofactor. This facilitates the synthesis of a protein attached to GFP and even allows for GFP expression in an entire organism (Lakowicz 2006; Kain et al. 1995). The protein does not fluoresce immediately after translation, but in about an hour, a result of self-catalyzed posttranslational modification, producing a fluorescent chromophore within the center of a barrel of eleven β-strands, as shown in Fig. 4.1 (Alberts et al. 2002). Additionally, it is stable; does

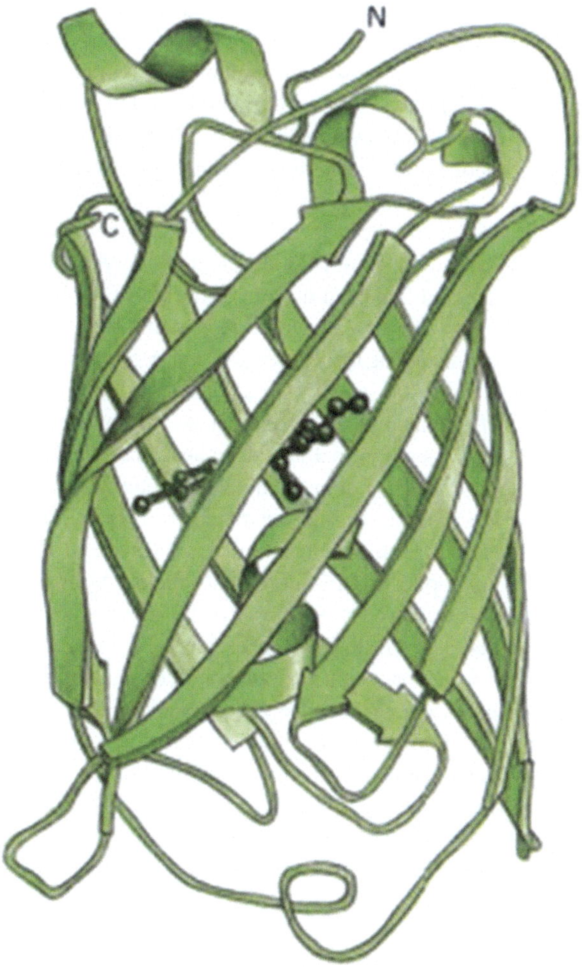

Fig. 4.1 A schematic diagram of GFP. The chromophore becomes active after posttranslational modifications. Adapted from Alberts et al. (2002)

not require a substrate, independent of the host species; and is considered to be relatively noninvasive in live cells (Hofmann 2010). GFP was first used in tracking bacterial gene expression, as well as in the sensory neurons of *C. elegans* (Chalfie et al. 1994). Since the introduction of GFP for scientific applications, there have been significant expansions in color ranges, as well as enhancements in the brightness and photostability (Shaner et al. 2008).

GFP can be used to monitor gene expression pattern by placing GFP-coding DNA sequence under the control of the promoter of a gene of interest (Alberts et al. 2002). Similarly, protein expression can also be monitored by fusing the GFP-coding sequence to the N or C-terminus to another protein, resulting in a chimeric protein and retaining the behaviour of the untagged, original protein, as the GFP would be controlled by the same regulatory elements. This can be demonstrated by using the GFP-tagged protein to rescue a mutant lacking the functional version of the protein. Therefore, it is possible to investigate the distribution and activities of the tagged protein in a living organism (Alberts et al. 2002). However, the high stability of GFP can also be a limitation. Temporal protein analysis can be hindered by "leaky" GFP expression of inducible systems that can contribute to a background fluorescence (van Roessel and Brand 2002). This can also be challenging in vivo, in which residual GFP can impede the expression history of a tissue. This can be attenuated by utilizing destabilized variants of GFP, but sensitivity may be compromised, as higher levels of protein are required to achieve a detectable signal (van Roessel and Brand 2002; Li et al. 1998). Although GFP is a well-studied fluorophore and has been used in a myriad of applications, the folding properties of GFP variants are not as well-understood. These molecules have presented challenges, especially at the expression temperature of 37 °C. YFP, in particular, has demonstrated to be sensitive to acidity and quenched by chloride ions (Cl^{-}) (Nagai et al. 2002). Nagai et al. developed an improved version of YFP, called "Venus," containing a mutation that is able to accelerate the chromophore oxidation, which is also the rate-limiting step to the YFP maturation (Nagai et al. 2002). As a result, Venus was found to fold relatively well, exhibiting an increase in speed and efficiency of maturation and higher resistance to the environmental changes; this enabled its application in fluorescent labeling with increased reliability (Nagai et al. 2002). Red fluorescent protein (RFP) is another common fluorescent protein from *Discosoma* sp. (DsRed) (Piatkevich and Verkhusha 2011). Another variant, mCherry RFP, has been used in a variety of quantitative imaging techniques, such as the fluorescence resonance energy transfer (FRET; as discussed below). With the emergence of GFP variants, multicolor live cell imaging is now possible that enables monitoring of several protein expression levels simultaneously (Higuchi-Sanabria et al. 2016; Shcherbakova and Verkhusha 2013).

GFP is not commonly accepted as a quantitative reporter of promoter activity in eukaryotic cells, but this can still be achieved by measuring the individual fluorescence by flow cytometry. Several studies have been conducted to explore the potential of GFP as a quantitative gene expression reporter at the single-cell level in eukaryotes. Soboleski et al. observed GFP fluorescence to be proportional to the GFP gene copy numbers when placed under the control of different eukaryotic promoters, including the (1) CMV immediate-early promoter, (2) doxycycline-inducible TRE promoter, and (3) VP1616-inducible HSV-1 immediate-early promoter (Soboleski et al. 2005). Therefore, it has been suggested that GFP can be used as a reliable, quantitative reporter of gene expression. GFP in bacterial systems can be used in expression vectors with lac promoters and the fluorescent protein sequences can be amplified by PCR, excised via the flanking restriction sites, then cloned into other expression vector of choice (Tangney and Francis 2012). GFP activity can then be assayed by simple plate counting, fluorescence and confocal microscopy, and flow cytometry—which can provide both qualitative and quantitative measurements (Tangney and Francis 2012).

Other applications of fluorescent proteins include molecular labels for studying localization and trafficking of specific organelles or recombinant proteins in live cells (Rizzuto et al. 1995). Localization studies of organelles can be achieved by fusing the gene for a fluorescent protein to a cDNA of a protein or peptide specific to an organelle of interest. Fluorescence microscopy enables the observation of morphology, dynamics, and distributions of the organelles over time (Rizzuto et al. 1995).

Protein interactions have been studied with fluorescent proteins via a technique known as FRET. Two molecules of interest are fluorescently labeled with two different fluorochromes; these fluorochromes will possess overlapping emission spectrum and the absorption spectrum of the other. When the two tagged proteins are within proximity of each other—a characteristic to protein interaction—there is a transfer of energy from one fluorochrome to another as a result of the excitation of the first fluorochrome. Therefore, by using GFP variants, proteins and the interaction with their receptors can be studied (Alberts et al. 2002).

Apoptosis can also be imaged in real time, via recombinant proteins. For example, recombinant photoprotein reporters that are activated by caspase-3 cleavage can effectively enable the visualization of dying cells (Bullok and Piwnica-Worms 2005). At the single-cell level, apoptosis can be imaged by labeling cells with fluorochromes in both the nucleus and the cytoplasm and observing the ratio of fluorescence of cytoplasm to nucleus (Cf/Nf) over time (Condeelis and Weissleder 2010; Mempel et al. 2006). Typically, dead cells display a decreased Cf/Nf ratio as a result of DNA fragmentation and loss of cell membrane integrity. Annexin V protein, a probe highly sensitive and specific to phosphatidylserine (PS), can be used to assay apoptosis. In dying cells, there is a redistribution of PS, in which PS gets flipped from the inner to the outer leaflet of the plasma membrane. With the GFP conjugation to Annexin V, apoptotic cells can be detected and sorted by using flow cytometry (Abbady et al. 2017).

GFP and other variants have been shown to be versatile tools in biology, as its genetic fusion to proteins of interest has allowed their visualization in living cells, and elicit their expression profiles and function.

4.1.2 Bioluminescence

Bioluminescence is a property found throughout the domains of life (Widder 2010). Although most of the species possessing the ability to emit light are aquatic, terrestrial examples can be found (Widder 2010; Shimomura 2006). The most notable being the common North American firefly (*Photinus pyralis*) (Fraga 2008). The main components in a firefly's production of light are luciferin, luciferase, ATP, and oxygen (Fraga 2008; McElroy et al. 1953). The reaction catalyzed by luciferase is a two-stage kinetic model (Fraga 2008; Deluca and McElroy 1978). The first stage is the conjugation of AMP with luciferin, whereas the second stage is the oxidation of the thiazoline ring found in luciferin (Fraga 2008; Deluca and McElroy 1974, 1978). This oxidation is the stage in which energy is released in the form of light in a waveband of around 560 nm, associated with a yellow-green color (Fraga 2008; Deluca and McElroy 1978; Seliger and McElroy 1960).

Although firefly luciferase (FLuc) is the most commonly used luciferase, it is not the only available option (Close et al. 2011). Another luciferase has been isolated from *Renilla reniformis*, a soft coral (Close et al. 2011; Lorenz et al. 1991). This luciferase is known as Renilla luciferase (RLuc), and although it produces light in a very similar reaction to FLuc, the light produced is slightly blue-shifted at a wavelength of 480 nm (Close et al. 2011; Lorenz et al. 1991). This makes RLuc less viable for examining tissue samples as more of the light is absorbed by the tissues and therefore less photons reach the sensors (Close et al. 2011; Contag and Ross 2002; Sadikot and Blackwell 2005; Lorenz et al. 1991). In order to get around this obstacle, there have been variants of RLuc created that are red-shifted (Close et al. 2011; Lorenz et al. 1991). A third well-studied luciferase can be extracted from bacteria (Close et al. 2011). This luciferase (Lux) has an optimal wavelength of 490 nm, placing it in between FLuc and RLuc in terms of absorption (Close et al. 2011). However, an interesting property of Lux is that it is a multi-subunit protein consisting of A, B, C, D, and E subunits that utilizes a very different mechanism of light production (Close et al. 2011; Rodriguez et al. 1985). Instead of using luciferin as a substrate, Lux produces a fatty aldehyde chain with the C, D, and E subunits, and said fatty aldehyde is oxidized to produce light (Close et al. 2011; Rodriguez et al. 1985). This means that Lux is much more self-sufficient than FLuc or RLuc which can be very useful depending on the application (Close et al. 2011). It is also possible to use different luciferases within the same subject as they have very different optimal wavelengths (Mezzanotte et al. 2017; Stacer et al. 2013).

Studying the mechanism by which organisms produce light has allowed for some very interesting applications. One such application is the use of bioluminescence imaging (BLI). BLI uses reporter genes that code for bioluminescent proteins (Contag and Ross 2002). Said reporter genes can be introduced to specific cell lines and implanted into the subject (Williams et al. 1991). These implanted cells produce light from within an organism and can be detected using several different types of receptors including charge-coupled device cameras (Contag and Ross 2002; Honigman et al. 2001). In mammals, the resolution for images taken through this technique is typically quite high; this is mostly due to a very high signal-to-background ratio (Sadikot and Blackwell 2005; Brogan et al. 2012). However, this is often limited to the organ or tissue level as it has been noted to have a lower spatial resolution than other imaging techniques (Sadikot and Blackwell 2005; Brogan et al. 2012).

One of the major advantages to BLI is that it is capable of capturing very high-quality images without the need to sacrifice animal subjects (Sadikot and Blackwell 2005). BLI can therefore allow for imaging of a disease as it progresses or for quantification of cellular activity (Sadikot and Blackwell 2005; Benaron et al. 1997). It is also a chemical reaction that is independent of light, which is uncommon in other imaging methods (Brogan et al. 2012). This, however, is a double-edged sword as BLI also requires injections of luciferin (for FLuc or RLuc) in order for the tissue under examination to produce light (Brogan et al. 2012). The reliance of external substrates may indicate that the amount of light produced may not be entirely representative of internal processes (Sadikot and Blackwell 2005; Rice et al. 2001).

The relatively recent use of BLI and its validation as an imaging technique has allowed bioluminescence to be used in many areas of research including gene monitoring, transgene delivery, and in the detection of certain cancers.

4.1.2.1 Bioluminescence in Monitoring Gene Expression

BLI has shown to be a very helpful concept in evaluating the expression of genes (Sato et al. 2004; Benaron et al. 1997). By conjugating the luciferase gene to a promoter region for a gene of interest, it is possible to get a good quantification of the activity of said gene through the amount of photons detected (Close et al. 2011; Sato et al. 2004). It is also possible to use this technology to better understand and implement other transgenes (Sato et al. 2004; Bhaumik et al. 2004). When delivering genes, it can be difficult to understand the long-term effects of a transgene (Sato et al. 2004). BLI can aid in the understanding of long-term effects through monitoring gene expression when luciferase is integrated with the gene of interest (Deroose et al. 2006). For example, luciferase activity has been shown to be a good indication of transgene expression delivered through a lentiviral vector in the central nervous system when both are linked together for delivery (Deroose et al. 2006).

4.1.2.2 Bioluminescence in Disease Progression

Knowing that bioluminescent proteins can be used to monitor activity of host and integrated transgenes, one can see an application for this technology in monitoring the progression of foreign DNA as seen in an infection (Sato et al. 2004). BLI can be used to access an infection from several perspectives including the integration of foreign DNA as well as the hosts immune response (Sadikot and Blackwell 2005; Sato et al. 2004). For example, in the course of pneumonia caused by *Pseudomonas aeruginosa*, it has been shown to be a good method of monitoring cytokine activation, specifically of NF-κB (Sadikot and Blackwell 2005; Sadikot et al. 2002). BLI has also proven its use for examining the route of infection as well as its progression through time (Sato et al. 2004). By transforming a strain of bacteria with a plasmid conferring bioluminescence or through direct integration of luciferase genes into the chromosome, it is possible to image the bacteria as it reproduces and spreads (Sato et al. 2004; Contag et al. 1995). The first example of this was done using *Salmonella*; however, it has been adapted for other bacteria and has been adapted for the visualization of viral infections as well (Sato et al. 2004; Contag et al. 1995).

4.1.2.3 Bioluminescence in Cancer Monitoring

BLI has been used with great promise in the field of oncology. Much like how BLI can track infection progression, it can also be used to show the progression of cancer and the efficiency of its treatment with gene therapy (Contag and Ross 2002; Wetterwald et al. 2002). Specifically, BLI has been shown to be an asset in assessing the metastasis of tumors (Contag and Ross 2002; Sadikot and Blackwell 2005; Sato et al. 2004; Wetterwald et al. 2002). Metastasis can be monitored through the development of tumor cells that express luciferase (Sato et al. 2004; Wetterwald et al. 2002). By introducing these engineered tumor cells into immunosuppressed animal models, it is possible to follow the pathogenesis of various cancers (Sadikot and Blackwell 2005; Wetterwald et al. 2002). BLI can also be used to help characterize and differentiate animal models of cancer based on their altered tumor suppressor genes, indicating the role that BLI can play in the understanding of cancer development and propagation as a whole (Sato et al. 2004).

4.1.2.4 Bioluminescence in Studying Protein-Protein Interactions

An interesting use of bioluminescence in research can be seen in the use of bioluminescence resonance energy transfer (BRET) (Boute et al. 2002). To fully understand BRET, it is best to examine where the inspiration has come from. BRET occurs naturally in *Renilla reniformis*, which is the source of RLuc as well as a fluorescent protein called Renilla green fluorescent protein (RGFP) (Boute et al. 2002; Morin and Hastings 1971). As stated, RLuc produces a wavelength of light at 480 nm (Close et al. 2011; Boute et al. 2002). In *Renilla* the two proteins are close enough so that RLuc can transfer energy to Renilla GFP; this energy transfer causes RGFP to produce light at a wavelength of 509 nm (Boute et al. 2002). The energy donor is usually RLuc, but the energy receiver can be any protein capable of fluorescence (Boute et al. 2002). This naturally occurring process can be repurposed to occur between any two proteins by conjugating one of the proteins with RLuc and the other with a fluorescent protein (Boute et al. 2001, 2002). Often, yellow fluorescent protein is used for the second protein as it is more able to penetrate the skin and has a higher contrast to RLuc than the RGFP (Close et al. 2011; Boute et al. 2002). If the proteins are close enough to interact, the wavelength produced by the fluorescent protein will be detected; if not the wavelength detected will be that of RLuc (Boute et al. 2001, 2002).

This technology is an adaptation of a similar technique known as fluorescence resonance energy transfer (FRET) (Boute et al. 2002). In FRET, both proteins are fluorescent, and the first is activated with an external light source (Boute et al. 2002). An obvious benefit of BRET is that it does not require this external light source; however, it has the drawback of being less well suited for modeling protein interactions on the subcellular level (Boute et al. 2002). BRET can be used for examining how naturally occurring proteins interact, but it may also be used to examine the state of receptors in terms of drug discovery (Boute et al. 2001, 2002).

4.1.2.5 Current and Future Advancements in Bioluminescence

Current areas of improvement in BLI include genetic engineering of red-shifted bioluminescent proteins as well as more red-shifted fluorescent proteins to be used in BRET (Mezzanotte et al. 2017). There has also been development and research in new luciferases such as NanoLuc, which has been isolated from a deep-sea shrimp and has the benefit of being only 19.1 kDa as compared to FLuc which is 62 kDa (Mezzanotte et al. 2017; Gould and Subramani 1988). Another current development is the visualization of calcium channels using photoproteins which are very similar to luciferases except that instead of a prolonged glow, they produce short flashes (Mezzanotte et al. 2017; Bakayan et al. 2015). This can allow for a better representation of rapidly activated and changing proteins (Mezzanotte et al. 2017; Bakayan et al. 2015). Current research suggests that there are some bacterial strains that grow preferentially on tumor cells; these can be transfected with luciferases in order to render a three-dimensional image of the cancer, which can lead to increased detection and possible treatment (Mezzanotte et al. 2017; Cronin et al. 2012).

Since the discovery and characterization of luciferase, the application of bioluminescence in research has grown dramatically. Although BLI is comparable in many applications to previously well-studied imaging techniques, such as PET, MRI, and radiography, it has still been underrepresented in the scientific literature (Contag and Ross 2002; Mezzanotte et al. 2017). Despite the originally low popularity of this imaging method, it has been shown to be a potential asset in the field of transgene delivery and in studying protein-protein interactions as well as for uses other than those aforementioned (Mezzanotte et al. 2017). Since the first discovery and characterization of bioluminescent proteins, their use in visualization studies has greatly expanded. The integration of bioluminescence as an imaging technology holds much exciting potential as it continues to develop and be refined.

4.1.3 MRI Reporter Genes

MRI reporter gene imaging offers several unique advantages including (1) the in vivo assessment of gene expression concurrent with acquisition of anatomic and functional MRI data; (2) the potential for serial temporal imaging in time-lapse studies that analyzes transient changes in gene expression throughout the gene delivery process; and (3) the investigation of dynamic processes associated with the cells of interest (e.g., disease progression or therapeutic effects) (Vandsburger et al. 2013). MRI reporter gene imaging involves the imaging of unique spectroscopic signatures through magnetic resonance spectroscopy (MRS) or the imaging of cells with enhanced contrast, due to the interaction between reporter gene expressions encoding a receptor/transporter with the enzyme with the administered contrast agent (Kang and Chung 2008). Creatine kinase gene has been used as a reporter gene through detection of ^{31}P MRS using phosphocreatine, the product of creatine kinase enzymatic reaction, as the indicator (Himmelreich and Dresselaers 2009). Other reporter genes suitable for imaging by MRS include cytosine deaminase and deoxycytidine kinase genes. There is great potential in using cytosine deaminase reporter genes as it can function as a therapeutic gene in targeted chemotherapy. The administration of non-cytotoxic 5-fluorocytosine prodrug, in conjunction with cytosine deaminase transgene expression, induces cell death through cytosine deaminase-mediated conversion of the prodrug into the active chemotherapeutic agent 5-fluorouracil (Waerzeggers et al. 2009). Thus, the multiple utility of cytosine deaminase permits targeted chemotherapy, as well as quantification of active 5-fluorouacil via ^{19}F MRS for studying pharmacokinetics and dose-response relationships.

MRI enzyme-based reporter gene imaging relies on the enzymatic action of the transgene product that leads to enhanced contrast. Human tyrosinase is an enzyme responsible for the series of reactions that result in melanin production. The overexpression of human tyrosinase from the transgene expression leads to elevated iron binding and enhanced MRI signals upon upregulated melanin production (Waerzeggers et al. 2009; Gilad et al. 2008; Hasegawa et al. 2010). However, human tyrosinase reporter gene has limited potential for clinical applications as iron binding by melanin constitutes the generation of high levels of reactive oxygen species (ROS) (Waerzeggers et al. 2009). The *LacZ* gene has been used as a MRI reporter gene through the enzymatic activity of beta-galactosidase cleaving (1-(2-(b-galactopyranosyloxy)propyl)-4,7,10-tris(carboxymethyl)-1,4,7,10-tetraazacyclododecane)gadolinium(III) (EgadMe) to expose the free coordination site of gadolinium for activation of MRI contrast (Kang and Chung 2008; Himmelreich and Dresselaers 2009; Waerzeggers et al. 2009; Gilad et al. 2008, 2009; Hasegawa et al. 2010; Liu et al. 2011).

MRI iron-based reporter gene imaging involves the accumulation of iron to generate detectable changes in MRI contrast. Such contrast can be controlled through the manipulation of administered iron concentrations; this enables real-time, sequential imaging of different transient cellular events, including gene expression (Gilad et al. 2008). Transferrin receptor, ferritin, and MagA genes are three iron-based reporter genes used in MRI reporter imaging; the overexpression of transferrin receptor results in elevated iron uptake through transferrin, a plasma protein (Gilad et al. 2008; Hasegawa et al. 2010). As a result, there is an elevated intracellular iron storage and subsequent iron accumulation sufficient for MRI contrast generation. The expression or co-expression of transferrin receptor and ferritin has been implicated as effective MRI reporters in many different applications. The bacterial counterpart for transferrin receptor and ferritin is the bacterial iron transporter, MagA, which serves to induce iron accumulation for enhanced contrast (Matsumoto and Jasanoff 2013). A novel method of MRI reporter imaging involves the transgene expression of artificial lysine-rich proteins (LRP) and subsequent contrast generation through chemical exchange saturation transfer (CEST) (Minn et al. 2015). CEST generates MRI contrast through selective labeling of protons on contrast agents using a radio-frequency saturation pulse and subsequent transfer of protons by chemical exchange agent with bulk water (Gilad et al. 2008; Hasegawa et al. 2010). With respect to LRP, CEST occurs upon the rapid transfer of labeled protons between amide functional groups and water (Liu et al. 2011). The application of LRP genes as MRI reporter genes is advantageous as LRPs are nonmetallic and biodegradable and can act as stand-alone contrast agent following the administration of radio-frequency saturation pulse (Gilad et al. 2009). Most importantly, CEST activation is switchable as contrast can only be detected when the applied saturation pulse frequency matches the properties of the exchangeable protons of contrast agents; this allows for simultaneous imaging of different CEST contrast agents possessing different excitation frequencies (Gilad et al. 2008).

MRI reporter genes have been recently introduced as a new platform for evaluation of gene expression; however, they have yet to be widely used due to several limitations. Low sensitivity of reporter genes at clinical magnetic field strength continues to be the major complication in MRI reporter imaging (Waerzeggers et al. 2009). In addition, the poor pharmacokinetics of administered contrast agents, the slow rates of iron accumulation, and the difficulties of interpreting MRI signal changes compound the issue and severely impact the evaluation of gene expression by MRI reporter gene systems (Liu et al. 2011; Gilad et al. 2009).

4.1.4 Radionuclide Reporter Imaging

Molecular imaging of positron emission tomography (PET) reporter gene expression offers a noninvasive means of in vivo monitoring of transgene expression; this allows for dynamic, real-time assessments of the entire gene delivery process (Youn and Hong 2012). Among the different image reporter genes suitable for various imaging techniques available (fluorescence, bioluminescence, MRI), radionuclide-based reporters remain the only reporter genes employed for noninvasive, whole-body imaging of transgene expression in humans (Yaghoubi et al. 2012). As PET provides qualitative and quantitative data with regard to signal intensity and regional/spatial distributions, the application of PET reporter gene/probe systems effectively quantifies transfection efficiencies and transgene expression levels (in correlation with signal intensity) while still providing critical information relating to site specificity and targeted delivery (Yaghoubi et al. 2012). PET reporter gene/probe systems involve transgene expression leading to the synthesis of a protein product capable of accumulating PET imaging probes within transfected cells (Fig. 4.2). PET probes may be enzyme substrates, ligands/antibodies, or transporter molecules for direct interactions with corresponding reporter gene product that result in elevated probe accumulation. PET reporter genes can encode (1) an enzyme that phosphorylates diffusible PET probes to alter its permeability across the cell membrane, (2) a protein receptor or antigen that binds to PET probes linked to specific ligands/antibodies, or (3) a cell membrane transporter that facilitates

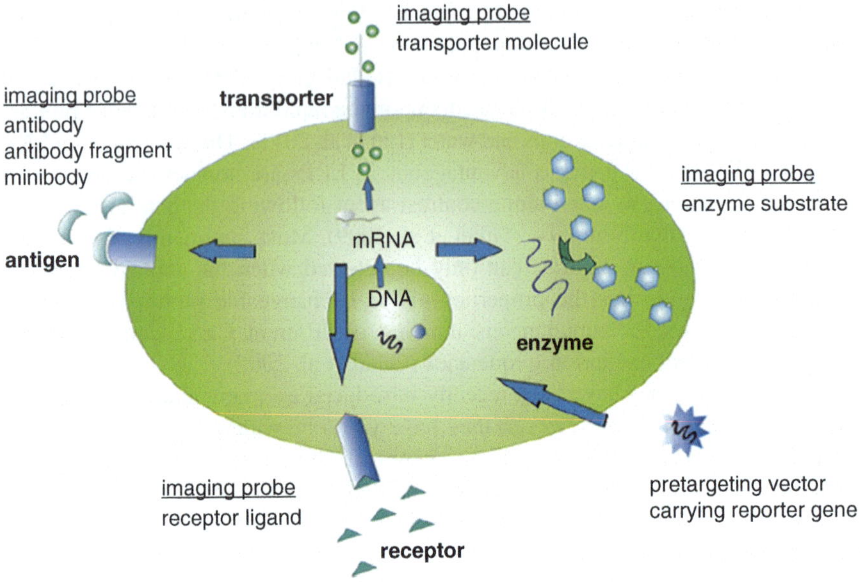

Fig. 4.2 Schematic of various imaging reporter gene/probe systems. Adapted from Hofmann (2010)

the entry and sequestration of PET probes within the cell (Yaghoubi et al. 2012; Waerzeggers et al. 2013).

The ideal PET reporter gene suitable for clinical gene therapy imaging would be capable of rapid, upregulated expressions, encode a highly targetable inert protein product that does not alter any cellular processes or trigger any immune responses, and generate image signals that directly correlates with the level of gene expression and protein production (Yaghoubi et al. 2012). Additionally, the corresponding PET probe must (1) be relatively stable for delivery to target cells, (2) be able to bypass the different physiological barriers prior to reaching the target cells, (3) accumulate only in cells expressing PET reporter genes, (4) be easily removed/metabolized (for background signal reduction and safety), and (5) possess suitable dosimetry in humans (suitable doses and repeat administrations) (Yaghoubi et al. 2012). For the purposes of evaluation, PET probes possessing radioisotopes with short half-life (^{11}C, ^{13}N, ^{15}O, ^{18}F) are more suitable over those with longer half-life (^{64}Cu and ^{124}I) as this permits repetitive imaging for detection of transient changes in gene expression following gene delivery (Waerzeggers et al. 2009).

Herpes simplex virus 1 thymidine kinase 1 (HSV1-TK) gene encodes an enzyme that phosphorylates target diffusible PET probes, composed of positron-emitting acycloguanosine or uracil derivatives, in order to intracellularly trap the probes within cells. This promotes PET probe accumulation and induces site-specific radioactivity corresponding to reporter gene expression (Liu et al. 2011; Yaghoubi et al. 2012; Serganova et al. 2008; Yaghoubi and Gambhir 2006; Jacobs et al. 2001).

The HSV1-TK reporter gene permits high sensitivity as each enzyme is capable of undergoing multiple phosphorylation reactions that result in signal amplification. The reporter gene also possesses high specificity as the PET probes involved in HSV1-TK reporter systems are poor substrates for mammalian thymidine kinase 1; hence, the membrane permeability of PET probes remains unaffected in non-transfected cells (Yaghoubi et al. 2012). Because HSV1-TK is the most extensively studied and applied PET reporter gene, numerous PET probes have been developed, possessing suitable pharmacokinetics in humans (Yaghoubi and Gambhir 2006). As result, the system has been implemented to image a variety of molecular events in living subjects (Jacobs et al. 2001; Massoud and Gambhir 2003; Wang et al. 2015; Alauddin and Gelovani 2010; Weissleder et al. 2000). The use of HSV1-TK gene possesses an additional advantage as it can function as the therapeutic gene in suicide gene therapy (Waerzeggers et al. 2013). The administration of ganciclovir, in conjunction with HSV1-TK transgene expression, induces cell death by inhibiting DNA synthesis via chain terminations upon phosphorylation by HSV1-TK enzyme. Thus, the HSV1-TK gene exhibits multiple utility and serves to function as an effective anticancer therapy while simultaneously monitoring of the effects of ganciclovir treatment on tumors. The prospects of HSV1-TK as a therapeutic and imaging modality have contributed to the generation of HSV1-TK mutants that increase sensitivity, reduce thymidine affinity, promote high affinity for specific PET probes, as well as shorten reporter gene half-life (Yaghoubi et al. 2012). Despite such improvements, the HSV1-TK reporter gene/probe system is limited by the inability of associated PET probes to cross the blood-brain barrier as well as the potential of immunostimulation that may arise from the viral gene (Yaghoubi and Gambhir 2006).

The issue of immunogenicity associated with HSV1-TK gene prompts the development of human gene-based PET reporters which effectively hinders the likelihood and severity of immune reactions. Numerous reporters have been explored and utilized in various studies as highlighted in a review by Yaghoubi et al. (2012). In particular, the sodium iodide symporter (NIS) gene is potentially an effective reporter gene as it possesses several key advantages: (1) limited endogenous expression (except thyroid and stomach) for reduced background signals, (2) rapid uptake and clearance of PET probes suitable for repetitive sequential imaging, and (3) availability of high-specificity and cost-effective IS-associated probes (Serganova et al. 2008). Similar to human somatostatin type 2 receptors, NIS has been previously utilized as a therapeutic gene in radionuclide therapy in addition to reporter gene imaging (Penheiter et al. 2012). The multiple utility of NIS gene has significant implications in its use in different clinical applications. Despite the potential of reduced immunogenicity, human-based PET reporters have several drawbacks, including the undesired accumulation of PET probes in cells naturally expressing the selected reporter gene and the alteration of cellular processes/characteristics due to the introduction of the reporter gene. These issues can be mediated through the induction of specific mutations that can eliminate or alter the inherent function of the reporter gene product, making them distinct from those endogenously expressed in cells (Yaghoubi et al. 2012). Such mutations may also serve to increase sensitivity through higher affinity with PET probes.

The incorporation of multiple reporter genes from different imaging modalities permit multimodal imaging that harnesses the strengths associated with each type of imaging technique. For example, concurrent imaging with MRI and PET combines the high 3D spatial resolution of MRI with high sensitivity of PET. Such multimodal imaging gathers clinically relevant anatomical, functional, and molecular data that cannot be obtained from each imaging modality on its own (Pichler et al. 2008). Ongoing efforts are focused into generating such multimodal, noninvasive imaging reporter genes for various applications. This has led to the development of a triple-fusion reporter vector, composed of reporter genes for bioluminescence, fluorescence, and PET, which possesses high activity for each respective protein constituent (Serganova et al. 2008).

PET reporter genes, as well as other reporter genes, can indirectly image therapeutic gene expression by tethering therapeutic and reporter gene expression. Linking reporter genes with therapeutic genes enables indirect quantification of therapeutic gene expression in addition to determining the therapeutic potential and corresponding optimal doses to yield therapeutic effects (Yaghoubi et al. 2012). The reporter gene can be linked to the therapeutic gene in five ways (Fig. 4.3): (1) development of an IRES-based singe gene expression system that encodes both the PET reporter gene and the therapeutic gene of interest, (2) co-administration of therapeutic and reporter gene delivery vectors possessing identical regulatory elements, (3) development of a therapeutic-reporter fusion gene under the regulation of a single promoter, (4) development of a therapeutic-reporter fusion gene under the regulation

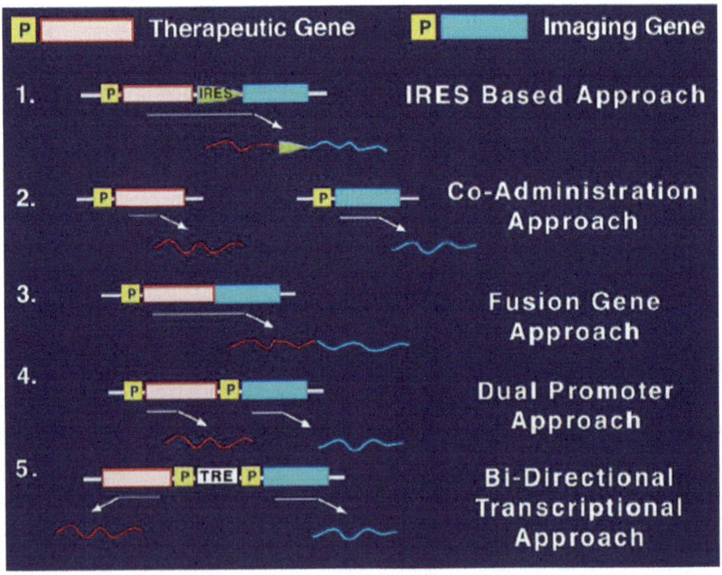

Fig. 4.3 Reporter gene mediation of indirect imaging of therapeutic genes. Adapted from Liu et al. (2015)

of two distinct promoters, and (5) development of a bidirectional gene expression system where both therapeutic and reporter genes are bidirectionally expressed under separate but identical minimal promoters (Yaghoubi et al. 2012).

4.2 Flow Cytometry and FACS

Flow cytometry is a sensitive cell analysis and cell sorting technique capable of rapidly measuring multiparameter characteristics of individual particles—most commonly cells, homogenously suspended in fluid during analysis (Davey and Kell 1996). A flow cytometer will illuminate fluorescent particles and cells as they pass a light source individually and emit signals (Givan 2011). Flow cytometry provides information to support studies in the biological systems including, but not limited to, cancer research, cell-based assays, cell signaling pathways, immune system network, and gene and drug delivery and development studies in a high-throughput fashion (Jahan-Tigh et al. 2012).

The fluorescence-activated cell sorter (FACS) is a specialized technology based on flow cytometry, designed as a powerful technique to analyze multiple parameters of an individual cell within a heterologous population. It offers a high level of purification for cell populations (often 99%) expressing limited differential markers (Jahan-Tigh et al. 2012). The collected data may then be further used to conduct other techniques to assess functional capability and status of cells. FACS machine is a high-speed cell sorter which measures and sorts fluorescently labeled cells (Herzenberg et al. 1976). These cells can be prokaryotic (e.g., bacteria) or eukaryotic (e.g., mammalian) requiring a minimum size of 0.2 μm (maximum 100 μm) (Robert et al. 2012). Flow cytometers are applied in a wide range of applications—from GFP expression analysis to cell counting and immunophenotyping. Flow cytometers function by passing thousands of cells, homogenously suspended in a buffer, within a brief second through a laser beam, and capturing the light emerging from the aligned cells as they sequentially pass through the sorting tube (Bonner et al. 1972). The data is gathered by the FACS software that statistically analyzes cellular characteristics such as size, surface, and intracellular protein expression, phenotype, health, and complexity. Hydrodynamic measurement is facilitated by a single-file alignment of cells through the laser beam (Herzenberg et al. 1976).

With the generation of fluorescent proteins as markers in genetically engineered cells and organisms, FACS have become an innovative tool in various fields of cell biology—from systemic biology and immunology to biopharmaceutical science. As a fluorescent cell passes through a laser beam, it would scatter light in different directions, in which the forward scatter (FS) would be detected and collected by the FACS detector (Givan 2011; Herzenberg et al. 1976; Bonner et al. 1972). The detector converts the intensity data collected from the scatter into a voltage pulse. The magnitude of the forward scatter is roughly proportional to the size of the cell, and a histogram of forward-scattered data graphically represents the size of the cells within a cell population (Herzenberg et al. 1976).

In addition to sorting cells based on size, cells can also be sorted based on the proteins expressed by the cell (Bonner et al. 1972). An antibody that is specific for the protein of interest is added to the cells, and these antibodies are then tagged with fluorochromes. The cells with the specific marker on the surface bound by antibodies will emit light under an excitation wavelength when the cells are interrogated by the laser. A cell emitting a particular wavelength can be collected as an electrical signal correlating to the number of fluorescent molecules is generated (Keith et al. 2000; Nafissi et al. 2014). Therefore, FACS functions to separate cells within preset ranges of both size and fluorescence, followed by deflection of the cell into an appropriate reservoir (Herzenberg et al. 1976).

One of the most common techniques to study vector characteristics using flow cytometer is to use fluorescent proteins such as GFP to investigate the efficacy of gene or drug delivery systems, as well as transfection efficiencies (Keith et al. 2000). A reporter construct with a specific promoter can express a fluorescent protein and then be introduced into cells; this permits the simultaneous measurement of the transfection efficiency and gene expression level (Keith et al. 2000). The fluorescent light emitted from transfected cells passing through the laser beam refracts in the same direction as the front-scattered signal. As the light passes through this tube, it passes through a series of filters, and particular wavelengths falling within preset ranges will be collected by the correlating detector. The fluorescent data will be collected as the population of transfected cells expressing GFP passes through the laser. As a result, a fluorescent signal will be generated and converted into a voltage pulse proportional to the amount of fluorescence emitted from the transfected cells, which can be represented on a histogram (Keith et al. 2000).

It is possible to run a double-fluorescent experiment with a single laser by using compatible fluorophores. For example, in our experiments, we used propidium iodide (PI) (red fluorescent) to detect the healthy population of transfected cells and the GFP (green fluorescent) to measure the transfected proportion of cells in a homogenous population (Nafissi et al. 2014). A dot plot is used to analyze the double-fluorescent data, and cells will be sorted into four distinct populations in the end, which will be represented in each quadrant in a dot plot. Cells only emitting bright red fluorescence will appear as dots on the upper left quadrant which represent dead cells (dead cells have lower forward scatter than living cells); cells only emitting bright green fluorescence will appear on the lower right quadrant, representing healthy cells that have been successfully transfected. Cells with both bright green and red emissions will appear on the upper right quadrant on the dot plot, which represent transfected but dead cells, and cells with low green and red signal will be shown on the lower left quadrant, which also represent the threshold. On a histogram, small cellular debris, autofluorescent cells, or media must be excluded from the data, which can be achieved by establishing a threshold, and this allows the majority of the data collected to solely consist of the cell population of interest (Herzenberg et al. 2006). For every FACS experiment, a threshold must be predefined, and only the data that exceed the threshold must be collected and analyzed. It is important to note that the data collected within the threshold (lower left) quadrant are not considered in the analysis. In our studies that utilized FACS, the GFP expression levels were calculated by multiplying the mean relative fluorescence values of transfected cells by the percentage of transfected cells. This parameter is considered to be directly proportional to the total amount of produced transgene product (Nafissi et al. 2014).

In a different experiment, we investigated the fate of the host cells after the foreign integration of a DNA vector into the host chromosome, in which we compared the number of early or late apoptotic cells to the healthy population of target cells (Nafissi and Slavcev 2012). Apoptosis was detected using the AnnexinV-FITC (ANN)/PI staining assay. Fluorescent ANN binds to plasma membrane-associated phosphatidylserine on the cell surface, and PI intercalates with DNA only by entering the dead cell through a broken plasma membrane. In this study, FACS data was used to indicate whether the integrated cells were healthy and viable, at early apoptotic stage, or dead.

In summary, flow cytometry is a unique advantageous tool that collects multiple parameters within the cell population in a complexed biological system and generates statistical data.

4.3 Reverse Transcription Polymerase Chain Reaction (RT-PCR)

One of the most important questions—if not the most important—in gene therapy is whether the genes of interest are expressed at the transcriptional and translational level. Reporter fusions are useful to assess expression and localization at the protein level, but do not offer much information at the RNA transcript level. Several techniques exist to evaluate the gene expression profile; among them, reverse transcription PCR (RT-PCR) techniques have been well standardized and have ultimately become the most common laboratory method of choice for the quantification of mRNA (Nolan et al. 2006). In principle, polymerase chain reaction (PCR) produces multiple copies of a particular DNA sequence through exponential amplification using temperature cycling to control in vitro polymerase activity on a DNA template fragment. In general, the cocktail for PCR requires a template (of the form DNA, RNA, cDNA), primers designed to amplify a target sequence (forward and reverse to capture the desired target), free nucleotides (dNTPs for DNA, NTPs for RNA), and a thermostable polymerase (Taq polymerase is a common example, derived from *Thermus aquaticus*) (Overbergh et al. 2003). Building upon this, RT-PCR utilizes the combination of reverse transcription and PCR. It is used to clone expressed genes by reverse transcribing the RNA of interest into its complement DNA (cDNA) through the use of reverse transcriptase (Overbergh et al. 2003). Subsequently, the newly synthesized cDNA is amplified using traditional PCR.

With the advent of real-time PCR, amplification of DNA can furthermore be detected in real time as the reaction progresses in contrast to standard endpoint PCR, where the product is only detected at its end (Nolan et al. 2006). This technique takes advantage of labeled probes at a number of different steps of the reaction, including but not limited to fluorogenic hybridization probes, fluorogenic hydrolysis probes, fluorogenic primers, and DNA binding dyes; based on the technique, RT-PCR can be quantitative (qPCR) or semiquantitative (Overbergh et al. 2003; Bustin et al. 2005; Wong and Medrano 2005). A key step in any RT-PCR experiment is normalization of gene expression to correct for variation across samples (Wong and Medrano 2005). Normalization is usually done against a control gene that is typically expressed at the same level regardless of experimental conditions. A control gene common across all tissue types remains highly elusive, if not absolutely impossible due to the high heterogeneity of cell types and experimental conditions (Andersen et al. 2004). Indeed, the appropriate reference gene would most likely differ between experiments. Aside from normalization, insufficient reporting of experimental protocols has raised concerns on RT-PCR data. Well-validated standard protocols must be agreed upon for robust quality control. To address this, the Minimum Information for Publication of Quantitative Real-Time PCR Experiments (MIQE) guidelines were established as a means of encouraging better experimental practice and standardizing PCR protocols for clearer data interpretation (Bustin et al. 2009). Real-time RT-PCR and qRT-PCR are far more informative over conventional PCR for the study of gene expression as they measure the amplification rate for one or more products; this can then be used to calculate the abundance of a target sequence as a measure of relative gene expression.

One famous example of RT-PCR is the TaqMan probes: short fluorescently labeled oligonucleotides that can be tracked during PCR cycles. First described in 1991 (Holland et al. 1991), these probes are now available commercially as the TaqMan® Gene Expression Assay (Thermo Fisher Scientific). Gene expression can be detected by observing the fluorescent activity of the probes. Hydrolysis of the TaqMan probes is accomplished in two steps, as outlined in Fig. 4.4: first, denaturation of the probe, which is then directly followed by annealing of the probe to the template, at which point the associated fluorescence emission can be measured (Nolan et al. 2006; Overbergh et al. 2003). The probes are dual-labeled with a reporter fluorochrome at the 5′ end and a quencher dye at the 3′ end. As the Taq polymerase synthesizes a new complementary strand of DNA from the template-bound probe, its intrinsic 5′–3′ exonuclease activity degrades the probe, thereby releasing the fluorophore and enabling fluorescence emission. Detection of the fluorescence upon each amplification cycle can then be used to determine the relative gene expression.

Various detection systems are used for real-time and RT-PCR analysis: (a) agarose gels, (b) fluorescent labeling of PCR products and detection with laser-induced fluorescence using capillary electrophoresis providing rapid detection of double-stranded DNA products (Williams et al. 1996; Richards and Poch 2002), and (c) plate capture and sandwich probe hybridization (Chou and Huang 2012). Other methods such as the Northern blot and the serial analysis of gene expression (SAGE) technique have been and still are used to analyze gene expression at the transcriptional (mRNA) level. James Alwine, David Kemp, and George Stark at Stanford

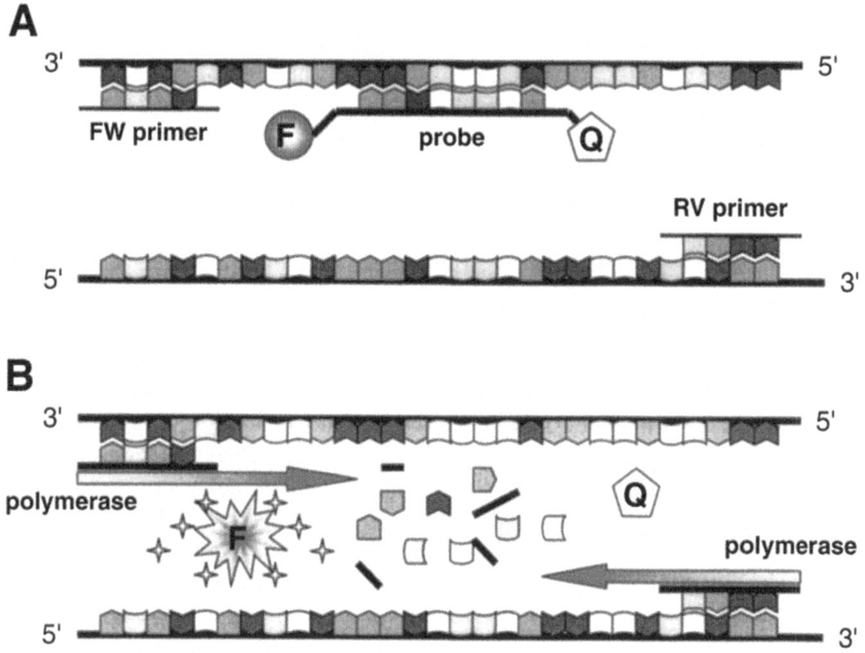

Fig. 4.4 Schematic representation of the TaqMan principle. (**a**) Probe and primers anneal to the target gene, but the probe remains intact, and fluorescence is not emitted. (**b**) The probe gets cleaved by the Taq polymerase during the extension phase, and fluorescence is emitted. *FW* forward, *RV* reverse, *F* fluorophore, and *Q* quencher dye. Adapted from Nolan et al. (2006)

University first described the Northern blot technique in 1977 (Alwine et al. 1977). In principle, the Northern blot enables detection of RNA through the hybridization of ^{32}P-labeled ssDNA/RNA probes to RNA that had been separated through electrophoresis and transferred to a suitable membrane for analysis. The Northern blot and related techniques can only detect known transcripts and are limited in their coverage.

Later, the SAGE technique was developed by Victor Velculescu, Lin Zhang, Bert Vogelstein, and Kenneth Kinzler in 1995 (Velculescu et al. 1995) in an effort to inspect the transcriptome, the collection of all RNA transcripts in an organism. SAGE is based on two principles: (1) even a very short nucleotide sequence (as short as 9 bp), referred to as a "tag," from an RNA transcript is sufficient to identify it; (2) sequencing is an expensive and time-consuming step so linking many short tags together into one fragment for sequencing drastically increases efficiency. Briefly, a sample of transcripts is reverse transcribed into cDNA and digested into very short fragments. These are then cocatenated into a longer string of tags and cloned in a vector for Sanger sequencing. Tag frequency is then a measure of transcript abundance, giving an idea of gene expression. SAGE is useful for assaying gene expression of unknown genes as it does not require any prior knowledge on transcript sequence.

Many more technologies have been developed to assay the transcriptome. Notably, DNA microarrays (reviewed extensively by Schulze and Downward 2001) were developed for high-throughput analysis of gene expression. Many short "probe" DNA sequences are immobilized on a surface (the array) and hybridized to target DNA or cDNA samples under highly stringent hybridizing conditions (Schulze and Downward 2001; Slonim and Yanai 2009). Alternatively, RNA sequencing (RNA-Seq) assesses the whole transcriptome at a given timepoint by utilizing deep sequencing techniques to sequence all transcripts in a biological sample (Wang et al. 2009). Similarly to MIQE, good practice guidelines have been established for microarray (Brazma et al. 2001) and RNA-Seq (T. E. Consortium 2011/2016) experiments.

Quantification and real-time monitoring of gene expression by way of monitoring RNA transcript levels have greatly advanced how biological activity can be measured. Protein-level fluorescent techniques enable the monitoring of protein localization and protein-protein interactions. Gene-based delivery of fluorescent and bioluminescent molecules has vastly improved our understanding of cellular structures and proteins and how they contribute to the overall biological system. However, a new challenging issue arising from expanding usage of modern transcriptomics technologies is one of data processing, interpretation, and statistical analysis. Such technologies grant far more data than ever before. There is, therefore, a great need not only for more sophisticated tools to handle the larger data output, but also for standardization across different laboratories in their toolset, data reporting methods, and statistical analyses.

References

Abbady AQ, Twair A, Ali B, Murad H (2017) Characterization of annexin V fusion with the superfolder GFP in liposomes binding and apoptosis detection. Front Physiol. https://doi.org/10.3389/fphys.2017.00317

Alauddin MM, Gelovani JG (2010) Radiolabeled nucleoside analogues for PET imaging of HSV1-tk gene expression. Curr Top Med Chem 10:1617–1632. https://doi.org/10.2174/156802610793176756

Alberts B, Johnson A, Lewis J et al (2002) Molecular biology of the cell, 4th edn. Garland Science, New York. https://doi.org/10.1091/mbc.E14-10-1437

Alwine JC, Kemp DJ, Stark GR (1977) Method for detection of specific RNAs in agarose gels by transfer to diazobenzyloxymethyl-paper and hybridization with DNA probes. Proc Natl Acad Sci U S A 74:5350–5354. https://doi.org/10.1073/pnas.74.12.5350

Andersen CL, Jensen JL, Ørntoft TF (2004) Normalization of real-time quantitative reverse transcription-PCR data: a model-based variance estimation approach to identify genes suited for normalization, applied to bladder and colon cancer data sets. Cancer Res 64:5245–5250. https://doi.org/10.1158/0008-5472.CAN-04-0496

Bakayan A, Domingo B, Miyawaki A, Llopis J (2015) Imaging Ca^{2+} activity in mammalian cells and zebrafish with a novel red-emitting aequorin variant. Pflugers Arch - Eur J Physiol 467 (9):2031–2042. https://doi.org/10.1007/s00424-014-1639-3

Benaron DA, Contag PR, Contag CH (1997) Imaging brain structure and function, infection and gene expression in the body using light. Philos Trans R Soc Lond Ser B Biol Sci 352 (1354):755–761. https://doi.org/10.1098/rstb.1997.0059

Bhaumik S, Lewis XZ, Gambhir SS (2004) Optical imaging of Renilla luciferase, synthetic Renilla luciferase, and firefly luciferase reporter gene expression in living mice. J Biomed Opt 9 (3):578–586. https://doi.org/10.1117/1.1647546

Bonner WA, Hulett HR, Sweet RG, Herzenberg LA (1972) Fluorescence activated cell sorting. Rev Sci Instrum 43:404–409. https://doi.org/10.1063/1.1685647

Boute N, Pernet K, Issad T (2001) Monitoring the activation state of the insulin receptor using bioluminescence resonance energy transfer. Mol Pharmacol 60(4):640–645. https://doi.org/10.1124/mol.106.026989.lin

Boute N, Jockers R, Issad T (2002) The use of resonance energy transfer in high-throughput screening: BRET versus FRET. Trends Pharmacol Sci 23:351–354. https://doi.org/10.1016/S0165-6147(02)02062-X

Brazma A, Hingamp P, Quackenbush J et al (2001) Minimum information about a microarray experiment (MIAME)—toward standards for microarray data. Nat Genet 29:365–371. https://doi.org/10.1038/ng1201-365

Brogan J, Li F, Li W et al (2012) Imaging molecular pathways: reporter genes. Radiat Res 177:508–513. https://doi.org/10.1667/RR2918.1

Bullok K, Piwnica-Worms D (2005) Synthesis and characterization of a small, membrane-permeant, caspase-activatable far-red fluorescent peptide for imaging apoptosis. J Med Chem 48:5404–5407. https://doi.org/10.1021/jm050008p

Bustin SA, Benes V, Nolan T, Pfaffl MW (2005) Quantitative real-time RT-PCR - a perspective. J Mol Endocrinol 34:597–601. https://doi.org/10.1677/jme.1.01755

Bustin SA, Benes V, Garson JA et al (2009) The MIQE guidelines: minimum information for publication of quantitative real-time PCR experiments. Clin Chem 55:611–622. https://doi.org/10.1373/clinchem.2008.112797

Chalfie M, Tu Y, Euskirchen G et al (1994) Green fluorescent protein as a marker for gene expression. Sci 263:802–805. https://doi.org/10.1126/science.8303295

Chou CC, Huang YH (2012) Nucleic acid sandwich hybridization assay with quantum dot-induced fluorescence resonance energy transfer for pathogen detection. Sensors (Switzerland) 12:16660–16672. https://doi.org/10.3390/s121216660

Close DM, Xu T, Sayler GS, Ripp S (2011) In vivo bioluminescent imaging (BLI): noninvasive visualization and interrogation of biological processes in living animals. Sensors 11:180–206

Condeelis J, Weissleder R (2010) In vivo imaging in cancer. Cold Spring Harb Perspect Biol. https://doi.org/10.1101/cshperspect.a003848

Contag CH, Contag PR, Mullins JI, Spilman SD, Stevenson DK, Benaron DA (1995) Photonic detection of bacterial pathogens in living hosts. Mol Microbiol 18(4):593–603. https://doi.org/10.1111/j.1365-2958.1995.mmi_18040593.x

Contag CH, Ross BD (2002) It's not just about anatomy: in vivo bioluminescence imaging as an eyepiece into biology. J Magn Reson Imag 16:378–387. https://doi.org/10.1002/jmri.10178

Cronin M, Akin AR, Collins SA, Meganck J, Kim JB, Baban CK et al (2012) High resolution in vivo bioluminescent imaging for the study of bacterial tumour targeting. PLoS One 7(1). https://doi.org/10.1371/journal.pone.0030940

Davey HM, Kell DB (1996) Flow cytometry and cell sorting of heterogeneous microbial populations: the importance of single-cell analyses. Microbiol Rev 60:641–696. https://doi.org/10.1016/j.mimet.2004.10.020

Deluca M, McElroy WD (1974) Kinetics of the firefly luciferase catalyzed reactions. Biochemistry 13(5):921–925. https://doi.org/10.1021/bi00702a015

Deluca M, McElroy WD (1978) [1] purification and properties of firefly luciferase. Methods Enzymol 57:3–15. https://doi.org/10.1016/0076-6879(78)57003-1

Deroose CM, Reumers V, Gijsbers R et al (2006) Noninvasive monitoring of long-term lentiviral vector-mediated gene expression in rodent brain with bioluminescence imaging. Mol Ther 14:423–431. https://doi.org/10.1016/j.ymthe.2006.05.007

Fraga H (2008) Firefly luminescence: a historical perspective and recent developments. Photochem Photobiol Sci 7:146–158. https://doi.org/10.1039/b719181b

Gilad AA, Ziv K, Mcmahon MT et al (2008) MRI reporter genes. J Nucl Med 49:1905–1908. https://doi.org/10.2967/jnumed.108.053520

Gilad AA, Van Laarhoven HWM, Mcmahon MT et al (2009) Feasibility of concurrent dual contrast enhancement using CEST contrast agents and superparamagnetic iron oxide particles. Magn Reson Med 61:970–974. https://doi.org/10.1002/mrm.21928

Givan A (2011) Flow cytometry: an introduction. Methods Mol Biol 699:1–29. https://doi.org/10.1007/978-1-61737-950-5_1

Gould SJ, Subramani S (1988) Firefly luciferase as a tool in molecular and cell biology. Anal Biochem 175:5–13. https://doi.org/10.1016/0003-2697(88)90353-3

Hasegawa S, Furukawa T, Saga T (2010) Molecular MR imaging of cancer gene therapy: ferritin transgene reporter takes the stage. Magn Reson Med Sci 9:37–47. https://doi.org/10.2463/mrms.9.37

Herzenberg LA, Sweet RG, Herzenberg LA (1976) Fluorescence-activated cell sorting. Sci Am 234:108–117. https://doi.org/10.1038/scientificamerican0376-108

Herzenberg LA, Tung J, Moore WA et al (2006) Interpreting flow cytometry data: a guide for the perplexed. Nat Immunol 7:681–685. https://doi.org/10.1038/ni0706-681

Higuchi-Sanabria R, Garcia EJ, Tomoiaga D et al (2016) Characterization of fluorescent proteins for three- and four-color live-cell imaging in *S. cerevisiae*. PLoS One. https://doi.org/10.1371/journal.pone.0146120

Himmelreich U, Dresselaers T (2009) Cell labeling and tracking for experimental models using magnetic resonance imaging. Methods 48:112–124. https://doi.org/10.1016/j.ymeth.2009.03.020

Hofmann A (2010) Spectroscopic techniques: I. Spectrophotometric techniques. In: Principles and techniques of biochemistry and molecular biology. Cambridge University Press, Cambridge, pp 477–521

Holland PM, Abramson RD, Watson R, Gelfand DH (1991) Detection of specific polymerase chain reaction product by utilizing the 5'----3' exonuclease activity of *Thermus aquaticus* DNA polymerase. Proc Natl Acad Sci U S A 88:7276–7280

Honigman A, Zeira E, Ohana P, Abramovitz R, Tavor E, Bar I et al (2001) Imaging transgene expression in live animals. Mol Ther 4(3):239–249. https://doi.org/10.1006/mthe.2001.0437

Jacobs A, Tjuvajev JG, Dubrovin M, et al (2001) Positron emission tomography-based imaging of transgene expression mediated by replication-conditional, oncolytic herpes simplex virus type 1 mutant vectors in vivo. Cancer Res 61:2983–2995. PubMed PMID: 11306477

Jahan-Tigh RR, Ryan C, Obermoser G, Schwarzenberger K (2012) Flow cytometry. J Invest Dermatol 132:1–6. https://doi.org/10.1038/jid.2012.282

Kain SR, Adams M, Kondepudi A et al (1995) Green fluorescent protein as a reporter of gene expression and protein localization. Biotechniques 19:650–655

Kang JH, Chung J-K (2008) Molecular-genetic imaging based on reporter gene expression. J Nucl Med 49:164S–179S

Keith M, Farrell P, Behie I, Behie L (2000) Use of flow cytometry to rapidly optimize the transfection of animal cells. Biotechniques 28:14–54

Lakowicz JR (2006) Fluorophores. In: Principles of fluoescence spectroscopy, 3rd edn. Springer, Berlin, pp 63–69

Li X, Zhao X, Fang Y et al (1998) Generation of destabilized green fluorescent protein as a transcription reporter. J Biol Chem 273:34970–34975. https://doi.org/10.1074/jbc.273.52.34970

Liu G, Swierczewska M, Niu G et al (2011) Molecular imaging of cell-based cancer immunotherapy. Mol Biosyst 7:993

Liu Z, Lavis LD, Betzig E (2015) Imaging live-cell dynamics and structure at the single-molecule level. Mol Cell 58:644. https://doi.org/10.1016/j.molcel.2015.02.033

Lorenz WW, McCann RO, Longiaru M, Cormier MJ (1991) Isolation and expression of a cDNA encoding Renilla reniformis luciferase. Proc Natl Acad Sci U S A 88(10):4438–4442. https://doi.org/10.1073/pnas.88.10.4438
Massoud TF, Gambhir SS (2003) Molecular imaging in living subjects: seeing fundamental biological processes in a new light. Genes Dev 17:545–580. https://doi.org/10.1101/gad.1047403
Matsumoto Y, Jasanoff A (2013) Metalloprotein-based MRI probes. FEBS Lett 587:1021–1029. https://doi.org/10.1016/j.febslet.2013.01.044
McElroy WD, Hastings JW, Coulombre J, Sonnenfeld V (1953) The mechanism of action of pyrophosphate in firefly luminescence. Arch Biochem Biophys 46(2):399–416. https://doi.org/10.1016/0003-9861(53)90211-2
Mempel TR, Pittet MJ, Khazaie K et al (2006) Regulatory T cells reversibly suppress cytotoxic T cell function independent of effector differentiation. Immunity 25:129–141. https://doi.org/10.1016/j.immuni.2006.04.015
Mezzanotte L, van t' Root M, Karatas H, et al (2017) In vivo molecular bioluminescence imaging: new tools and applications. Trends Biotechnol 35:640–652. https://doi.org/10.1016/j.tibtech.2017.03.012
Minn I, Bar-Shir A, Yarlagadda K et al (2015) Tumor-specific expression and detection of a CEST reporter gene. Magn Reson Med 74:544–549. https://doi.org/10.1002/mrm.25748
Morin JG, Hastings JW (1971) Energy transfer in a bioluminescent system. J Cell Physiol 77 (3):313–318. https://doi.org/10.1002/jcp.1040770305
Nafissi N, Slavcev R (2012) Construction and characterization of an in-vivo linear covalently closed DNA vector production system. Microb Cell Fact 11:154. https://doi.org/10.1186/1475-2859-11-154
Nafissi N, Alqawlaq S, Lee EA et al (2014) DNA ministrings: highly safe and effective gene delivery vectors. Mol Ther Nucleic Acids 3:e165. https://doi.org/10.1038/mtna.2014.16
Nagai T, Ibata K, Park ES et al (2002) A variant of yellow fluorescent protein with fast and efficient maturation for cell-biological applications. Nat Biotechnol 20:87–90. https://doi.org/10.1038/nbt0102-87
Nolan T, Hands RE, S a B (2006) Quantification of mRNA using real-time RT-PCR. Nat Protoc 1:1559–1582. https://doi.org/10.1038/nprot.2006.236
Overbergh L, Giulietti A, Valckx D, et al (2003) The use of real-time reverse transcriptase PCR for the quantification of cytokine gene expression. J Biomol Tech 14:33–43. PMC2279895
Penheiter AR, Russell SJ, Carlson SK (2012) The sodium iodide symporter (NIS) as an imaging reporter for gene, viral, and cell-based therapies. Curr Gene Ther 12:33–47. https://doi.org/10.2174/156652312799789235
Permyakov E (1993) Luminescent spectroscopy of proteins. CRC Press, London
Piatkevich KD, Verkhusha VV (2011) Guide to red fluorescent proteins and biosensors for flow cytometry. Methods Cell Biol. https://doi.org/10.1016/B978-0-12-374912-3.00017-1
Pichler BJ, Judenhofer MS, Pfannenberg C (2008) Multimodal imaging approaches: PET/CT and PET/MRI. Handb Exp Pharmacol 185:109–132. https://doi.org/10.1007/978-3-540-72718-7-6
Rice BW, Cable MD, Nelson MB (2001) In vivo imaging of light-emitting probes. J Biomed Opt 6 (4):432–440. https://doi.org/10.1117/1.1413210
Richards MP, Poch SM (2002) Quantitative analysis of gene expression by reverse transcription polymerase chain reaction and capillary electrophoresis with laser-induced fluorescence detection. Mol Biotechnol 21:19–37. https://doi.org/10.1385/MB:21:1:019
Rizzuto R, Brini M, Pizzo P et al (1995) Chimeric green fluorescent protein as a tool for visualizing subcellular organelles in living cells. Curr Biol 5:635–642. https://doi.org/10.1016/S0960-9822(95)00128-X
Robert S, Lacroix R, Poncelet P et al (2012) High-sensitivity flow cytometry provides access to standardized measurement of small-size microparticles-brief report. Arterioscler Thromb Vasc Biol 32:1054–1058. https://doi.org/10.1161/ATVBAHA.111.244616
Rodriguez A, Nabi IR, Meighen E (1985) ATP turnover by the fatty acid reductase complex of Photobacterium phosphoreum. Can J Biochem Cell Biol 63(10):1106–1111

Sadikot RT, Wudel JL, Jansen DE, Debelak JP, Yull FE, Christonan JW et al (2002) Hepatic cryoablation-induced multisystem injury: bioluminescent detection of NF-κB activation in a transgenic mouse model. J Gastrointest Surg 6(2):264–270. https://doi.org/10.1016/S1091-255X(01)00064-6

Sadikot RT, Blackwell TS (2005) Bioluminescence imaging. Proc Am Thorac Soc 2:511–540. doi: https://doi.org/10.1513/pats.200507-067DS

Sato A, Klaunberg B, Tolwani R (2004) In vivo bioluminescence imaging. Comp Med 54:631–634

Schulze A, Downward J (2001) Navigating gene expression using microarrays—a technology review. Nat Cell Biol 3:E190–E195. https://doi.org/10.1038/35087138

Seliger HH, McElroy WD (1960) Spectral emission and quantum yield of firefly bioluminescence. Arch Biochem Biophys 88(1):136–141. https://doi.org/10.1016/0003-9861(60)90208-3

Serganova I, Mayer-Kukuck P, Huang R, Blasberg R (2008) Molecular imaging: reporter gene imaging. Handb Exp Pharmacol 185:167–223. https://doi.org/10.1007/978-3-540-77496-9-8

Shaner NC, Lin MZ, McKeown MR et al (2008) Improving the photostability of bright monomeric orange and red fluorescent proteins. Nat Methods 5:545–551. https://doi.org/10.1038/nmeth.1209

Shcherbakova DM, Verkhusha VV (2013) Near-infrared fluorescent proteins for multicolor in vivo imaging. Nat Methods 10:751–754. https://doi.org/10.1038/nmeth.2521

Shimomura O (2006) Bioluminescence, Chemical principles and methods, pp 1–29. https://doi.org/10.1142/9789812773647

Slonim DK, Yanai I (2009) Getting started in gene expression microarray analysis. PLoS Comput Biol 5:e1000543. https://doi.org/10.1371/journal.pcbi.1000543

Soboleski MR, Oaks J, Halford WP (2005) Green fluorescent protein is a quantitative reporter of gene expression in individual eukaryotic cells. Fed Am Soc Exp Biol 19:440–442. https://doi.org/10.1096/fj.04-3180fje

Stacer AC, Nyati S, Moudgil P, Iyengar R, Luker KE, Rehemtulla A, Luker GD (2013) NanoLuc reporter for dual luciferase imaging in living animals. Mol Imaging 12(7). https://doi.org/10.2310/7290.2013.00062

Tangney M, Francis KP (2012) In vivo optical imaging in gene & cell therapy. Curr Gene Ther 12:2–11. https://doi.org/10.2174/156652312799789299

T.E. Consortium (2011/2016) Standards, guidelines and best practices for RNA-Seq V1.0 (June 2011). https://genome.ucsc.edu/ENCODE/protocols/dataStandards/ENCODE_RNAseq_Standards_V1.0.pdf

van Roessel P, Brand AH (2002) Imaging into the future: visualizing gene expression and protein interactions with fluorescent proteins. Nat Cell Biol 4:E15–E20. https://doi.org/10.1038/ncb0102-e15

Vandsburger MH, Radoul M, Cohen B, Neeman M (2013) MRI reporter genes: applications for imaging of cell survival, proliferation, migration and differentiation. NMR Biomed 26:872–884. https://doi.org/10.1002/nbm.2869

Velculescu VE, Zhang L, Vogelstein B, Kinzler KW (1995) Serial analysis of gene expression. Science 270:484–487

Waerzeggers Y, Monfared P, Viel T et al (2009) Methods to monitor gene therapy with molecular imaging. Methods 48:146–160

Waerzeggers Y, Viel T, Schäfers S et al (2013) Positron emission tomography-based molecular imaging. Neuromethods 77:301–327. https://doi.org/10.1007/7657_2012_36

Wang Z, Gerstein M, Snyder M (2009) RNA-Seq: a revolutionary tool for transcriptomics. Nat Rev Genet 10:57–63. https://doi.org/10.1038/nrg2484

Wang Z, Wang F, Hida N et al (2015) Design of a functional cyclic HSV1-TK reporter and its application to PET imaging of apoptosis. Nat Protoc 10:807–821. https://doi.org/10.1038/nprot.2015.048

Weissleder R, Moore a MU et al (2000) In vivo magnetic resonance imaging of transgene expression. Nat Med 6:351–355. https://doi.org/10.1038/73219

Wetterwald A, van der Pluijm G, Que I, Sijmons B, Buijs J, Karperien M, Löwik CW, Gautschi E, Thalmann GN, Cecchini MG (2002) Optical imaging of cancer metastasis to bone marrow: a mouse model of minimal residual disease. Am J Pathol 160(3):1143–1153. https://doi.org/10.1016/S0002-9440(10)64934-6

Widder EA (2010) Bioluminescence in the ocean: origins of biological, chemical, and ecological diversity. Science (80–) 704:704–708. doi: https://doi.org/10.1126/science.1174269

Williams RS, Johnston SA, Riedy M et al (1991) Introduction of foreign genes into tissues of living mice by DNA-coated microprojectiles. Proc Natl Acad Sci U S A 88:2726–2730. https://doi.org/10.1073/PNAS.88.7.2726

Williams SJ, Schwer C, Krishnarao AS et al (1996) Quantitative competitive polymerase chain reaction: analysis of amplified products of the HIV-1 gag gene by capillary electrophoresis with laser-induced fluorescence detection. Anal Biochem 236:146–152. https://doi.org/10.1006/abio.1996.0143

Wong ML, Medrano JF (2005) Real-time PCR for mRNA quantitation. BioTechniques 39(1):75–85

Yaghoubi SS, Gambhir SS (2006) PET imaging of herpes simplex virus type 1 thymidine kinase (HSV1-tk) or mutant HSV1-sr39tk reporter gene expression in mice and humans using [18F] FHBG. Nat Protoc 1:3069–3075. https://doi.org/10.1038/nprot.2006.459

Yaghoubi SS, Campbell DO, Radu CG, Czernin J (2012) Positron emission tomography reporter genes and reporter probes: gene and cell therapy applications. Theranostics 2:374–391. https://doi.org/10.7150/thno.3677

Youn H, Hong KJ (2012) In vivo noninvasive small animal molecular imaging. Osong Public Health Res Perspect 3:48–59. https://doi.org/10.1016/j.phrp.2012.02.002

Chapter 5
Conclusions and Perspectives

Roderick Adrian Slavcev

The dawn of nanoparticle-encapsulated genes is a revolutionary move in gene therapy. It promises to specifically and safely transport genetic cargo through biological systems within a non-viral "Trojan horse" system. Advances in non-viral-based physical (sonoporation, electroporation) and chemical (calcium phosphate) methods surmount biological barriers for a more efficient gene delivery and gene expression. This decade has witnessed lots of tailor-made gene delivery complexes and their evaluation. For example, magnetofection and sonoporation can be combined through the coadministration of magnetic nanoparticles and nonmagnetic microbubbles, followed by their activation by applying magnetic field and ultrasound. Recent hybrid nanocomplexes are not only attractive systems for gene delivery but also serve as nanoprobes for cellular fluorescent, PET, and MRI imaging and for gene evaluation by fluorescence, bioluminescence, or by radionucleotide reporter methods. There is a plethora of techniques available, and one can choose from radioactive-based to nonradioactive-based evaluation methods. Technological advances in imaging techniques assure better resolution and sensitivity; for example, with the advent of two-photon multiphoton microscopy over the traditional confocal microscopy, it is now able to image sixfold deeper than confocal microscopy when imaging the same sample using the same fluorophores. In the future, one needs to take advantage of these rapidly progressing imaging techniques and evaluation tools for maximum gene delivery using newer and safer targeted forms of non-viral systems. These nanosystems promise to be of both diagnostic and therapeutic value in the same breath.

R. A. Slavcev (✉)
University of Waterloo, Waterloo, ON, Canada
e-mail: slavcev@uwaterloo.ca

© Springer International Publishing AG, part of Springer Nature 2018

R. A. Slavcev et al. (eds.), *Nanomedicine*, Experientia Supplementum 110,
https://doi.org/10.1007/978-3-319-78259-1_5

The manufacturer's authorised representative in the EU is Springer Nature Customer Service Centre GmbH, Europaplatz 3, 69115 Heidelberg, Germany. If you have any concerns regarding our products, please contact ProductSafety@springernature.com

Printed and bound by CPI Group (UK) Ltd, Croydon, CR0 4YY
05/10/2026
02228162-0001